BOOKLET MAKING

IN THE

CLASSROOM

by
Elaine Commins

illustrated by
Margherita De Paulis

Publishers
T.S. Denison and Company, Inc.
Minneapolis, Minnesota 55431

Standard Book Number: 513-02063-2
Booklet Making in the Classroom

Minneapolis, Minnesota 55431

INTRODUCTION

Booklet making is fun! Children enjoy the results of their own labors, and through booklet making projects they have the opportunity to create while they are learning. The final result is a product that can be taken home with pride.

There are various methods for preparing booklets that are included in this volume. Aside from differences in physical construction, a variety of content ideas are offered here which will enhance curriculum subjects. All suggestions for games, exercises and activities are designed to require a minimum of teacher supervision once the basic directions have been presented to students.

The various booklet themes in this volume fall under four major subject headings: pre-reading, reading and writing; mathematics; science; and social studies. Suggested booklet activities based on these subjects are designed for students in prekindergarten through grade 3.

Unless indicated **Group Booklet**, all booklets are designed as individual projects.

Table of Contents

Chapter 1

HOW TO CONSTRUCT BOOKLETS

Included in this chapter are various techniques employing:

- long-reach staplers
- desk-top staplers
- hole-punchers and yarn
- brads
- a binding machine

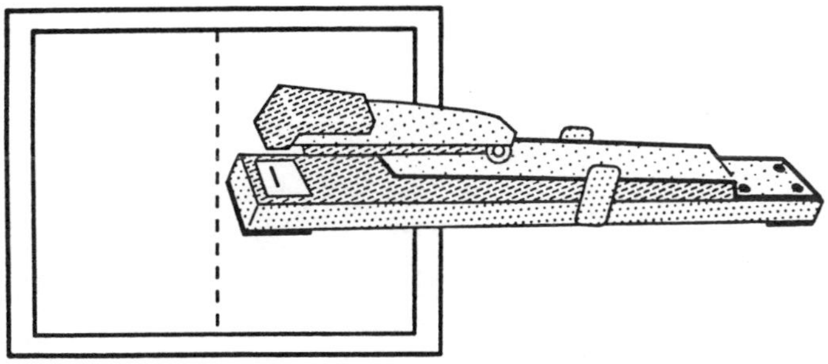

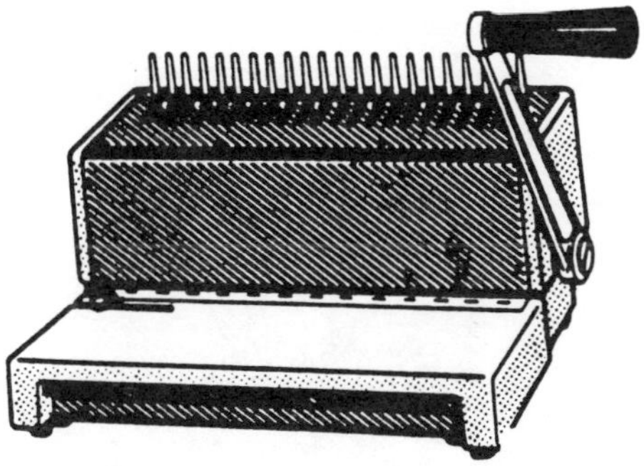

Every teacher searches for new ideas that make teaching easier. The most desirable ideas are those that foster independent work by students. One of the most proven and successful techniques is the preparation of booklets. Through booklet making, children can become involved in a hands-on approach which promotes self-teaching and creativity along with intellectual accomplishment.

Since booklet making can be adapted to almost any subject in the curriculum, it is important to vary the types of booklets used in order to sustain high interest. Included in this chapter are descriptions of several different styles of booklets. In the remaining chapters, various methods and ideas for their implementation in subject areas are discussed.

Most booklets contain only two elements: a cover and inner pages. The cover usually consists of colored construction paper that is folded in half. Substitutions such as oak tag, poster board or cloth-covered cardboard may be used instead of construction paper. The inner pages are often manila paper, typing paper, lined notebook paper, or worksheets that are inserted in the cover and attached with either staples, yarn, or brads. The following examples illustrate a variety of booklets that can be made in the classroom using inexpensive material.

1. **Using long-reach staplers**

   This is the easiest and fastest way to assemble a booklet. Using a 9" x 12" sheet of colored construction paper, two or more sheets of 8 1/2" x 11" typing paper are centered on the piece of construction paper and then stapled down the middle.

   When folded at the staple line, a booklet is formed.

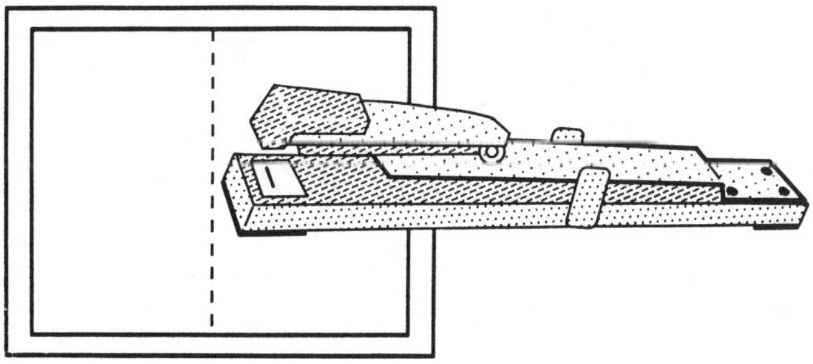

2. Using desk-top staplers

Another easy method uses regulation desk-top staplers.

A sheet of 12" x 18" construction paper is folded in half and used as a cover. Then a piece of 8 1/2" x 11" typing paper or 9" x 12" manila paper is inserted inside the cover and stapled to the back page.

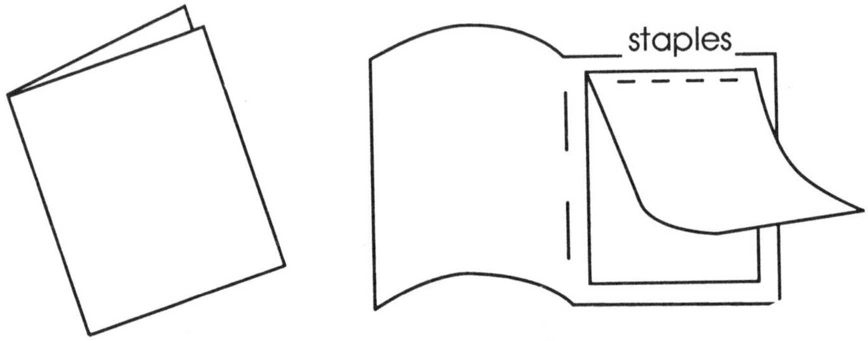

This same kind of booklet can be made by using a sheet of 9" x 12" construction paper for the cover and typing paper cut in half for the inside pages.

3. Using desk-top staplers

The cover consists of a sheet of 12" x 18" colored construction paper. It is folded in half and 8 1/2 " x 11" typing paper is inserted close to the fold. It is then stapled along the fold.

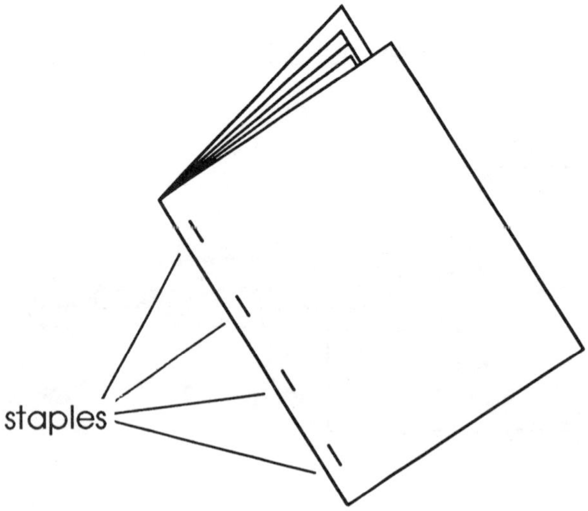

4. Using hole punchers and yarn

The cover, either a sheet of 9" x 12" or 12" x 18" colored construction paper, is folded and the inside pages are inserted close to the fold.

Using a hole puncher, two holes are punched along the folded edge, making sure the inside pages are included in the punch. One hole should be about one-quarter from the top and the second hole about one-quarter from the bottom. Yarn is inserted in the holes and the booklet is then tied together.

5. Using brads

Brads are usually used for group booklets and it is recommended that oak tag be used for the cover because it is more durable than construction paper. If construction paper is used, however, it should be laminated.

Two pieces of 13" x 19" oak tag are used as cover pages. The inside pages consist of 12" x 18" manila paper.

Brads are used to hold the cover and pages together.

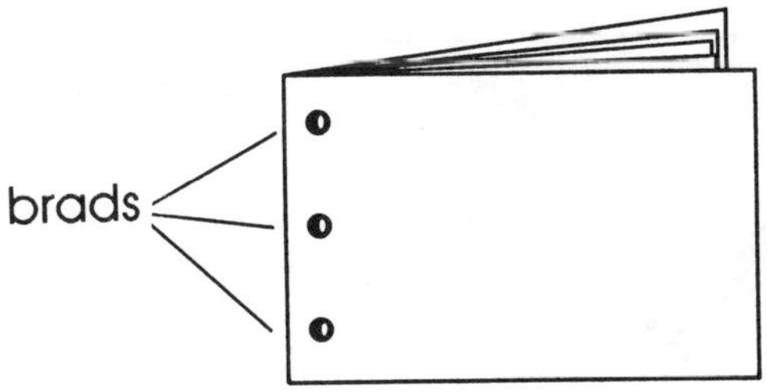

a brad

6. **Using desk-top staplers**

This is essentially a coverless booklet. Four or more pieces of construction paper or typing paper or a combination of the two are stapled together at the short end.

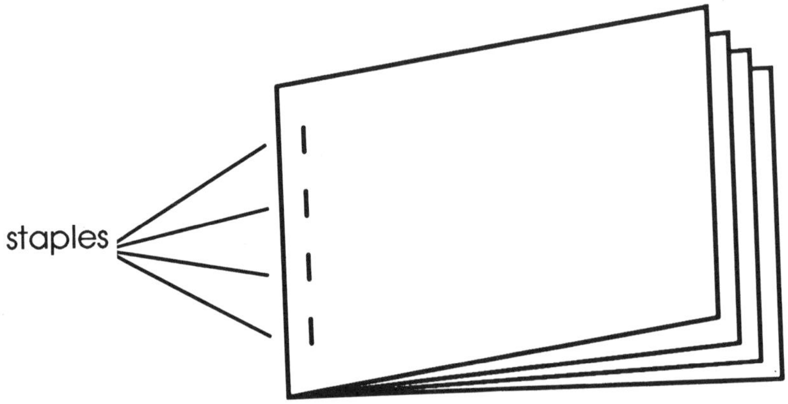

7. **Novelty booklets that use various fasteners**

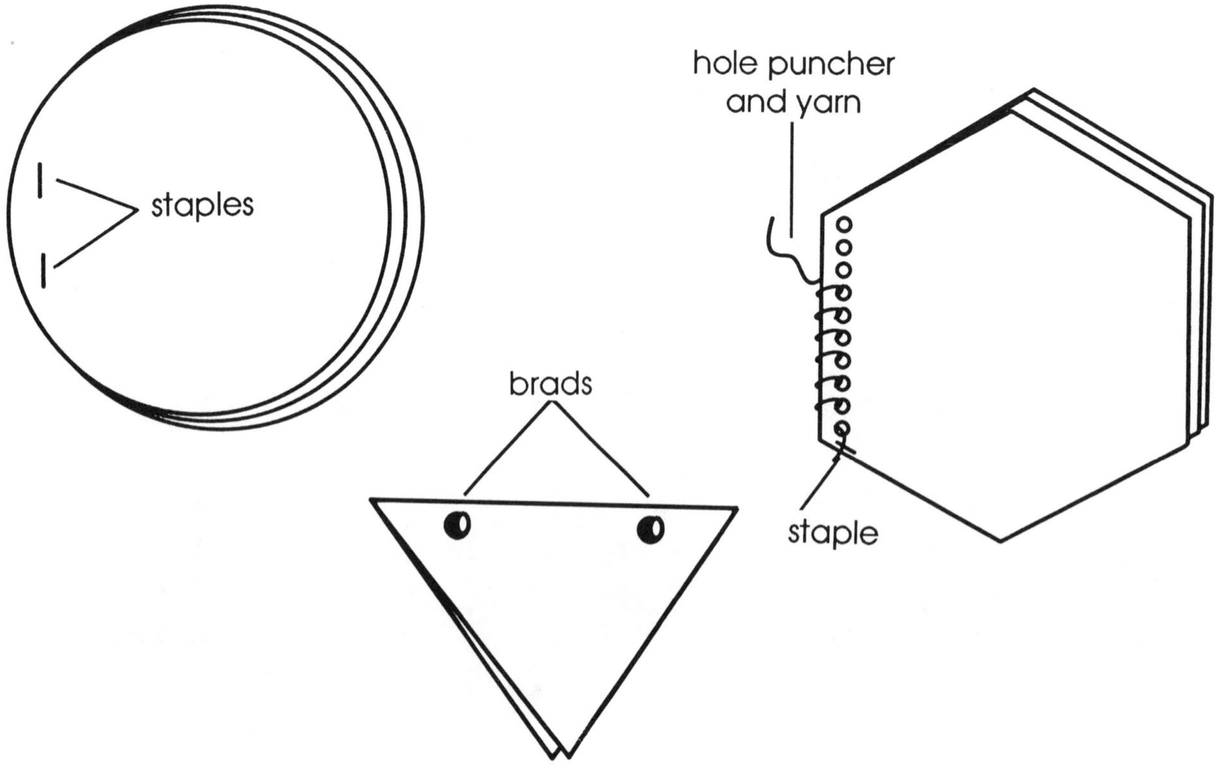

8. Many schools or school systems have binding machines that punch holes and insert plastic spiral strips that give booklets a professional look. If teachers have access to binding machines, they may use them often or only for special occasions such as holiday booklets, etc.

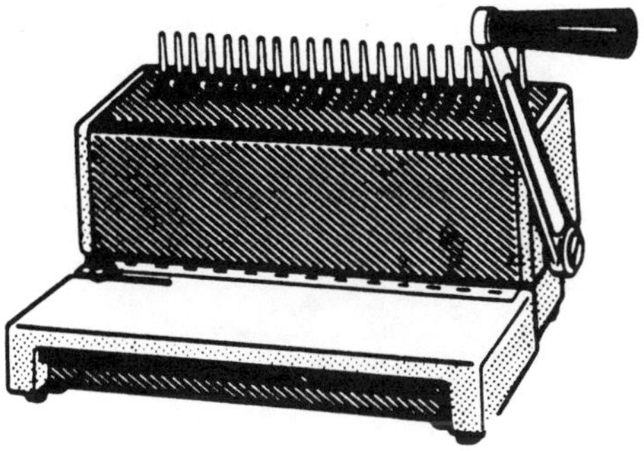

Note: Students usually decorate their own covers.

Chapter 2

PRE-READING, READING and WRITING

The activities in this chapter are planned to enhance the following skills:

- matching colors, shapes, letters and words
- rhyming
- sequencing
- describing
- using verbs
- identifying vowels
- rearranging letters to form words
- illustrating poems and stories
- writing poems, stories and book reports

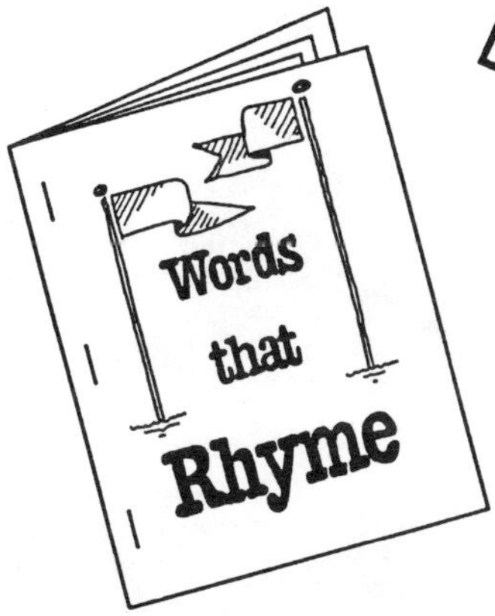

COLORS

Description of Booklet

This booklet consists of a 12" x 18" cover of colored construction paper that is folded in half. Inner pages of typing paper or manila paper are stapled inside the cover along one side.

Learning Skills Involved

Eye-hand coordination and visual discrimination.

Method of Preparation

Booklets are constructed in advance by the teacher. The name of a different color is printed at the top of each page. Colored crayons are used to print the names.

Children either draw a picture emphasizing the suggested color or they may search through magazines for objects of the color, cut them out, and paste them on the matching page.

Preparation Time

From one day to a week, depending on the number of colors used.

WRITING

Description of Booklet

This booklet consists of a 12" x 18" cover of colored construction paper that is folded in half. Several prepared worksheets are stapled inside the cover.

Learning Skills Involved

Eye-hand coordination.

Method of Preparation

Children may complete one worksheet each day which the teacher collects and saves. After four or five worksheets have been completed, they are stapled or braded into booklet form.

Covers may be decorated.

Preparation Time

Approximately five days.

Suggestions for worksheets are given on the next page.

WRITING

Suggestions for teacher-made worksheets:

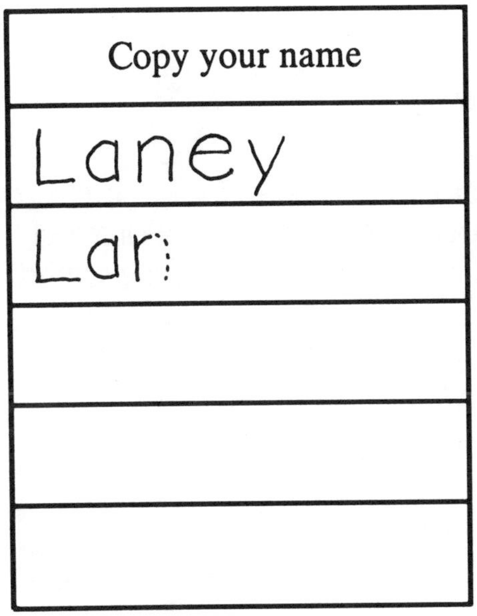

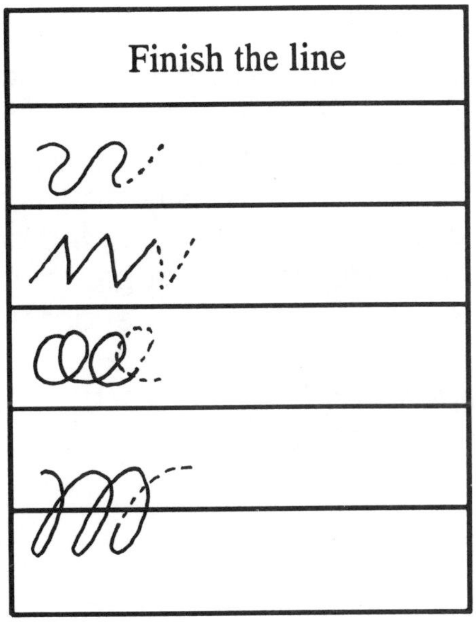

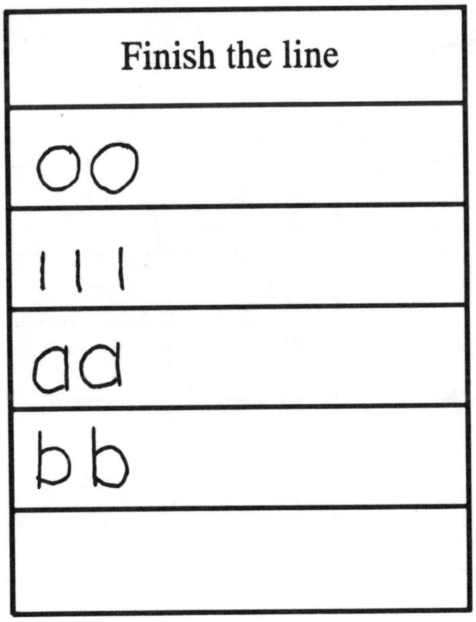

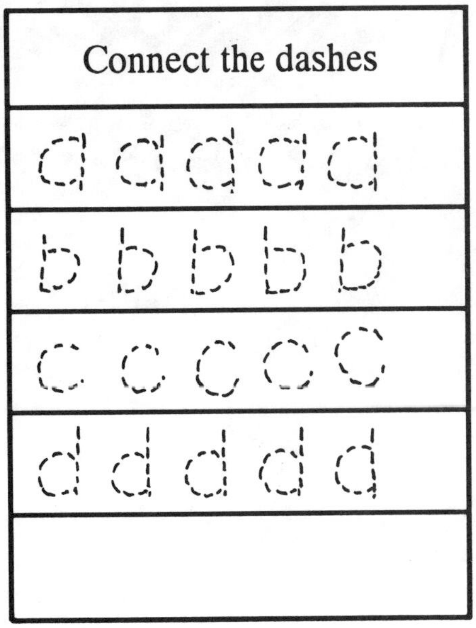

MY AUTOGRAPH BOOK

Description of Booklet

This booklet consists of a 9" x 12" cover of colored construction paper that is folded in half. Lined inner pages are stapled inside the back cover.

Learning Skills Involved

Eye-hand coordination, visual discrimination and collecting.

Method of Preparation

Booklets are constructed in advance by the teacher. Inner pages are stapled to the inside of the back cover page. Children walk around the room collecting autographs from their fellow students. First names only are suggested for younger children.

Preparation Time

One or two days.

MY ABC BOOK

Description of Booklet

This booklet consists of a 12" x 18" cover of colored construction paper that is folded in half. Inner pages of typing paper will be stapled inside the back cover when completed by students.

Learning Skills Involved

Eye-hand coordination and visual discrimination.

Method of Preparation

Children are given pencils and wooden alphabet letters or stencils to trace on typing paper. They may fill as many papers as they desire with letters. Letters may be colored with crayons. As each paper is completed, it is collected by the teacher who saves them and staples them into a booklet.

Older children might be asked to copy letters in alphabetical order. A written alphabet may be placed on the work table for handy reference.

Preparation Time

One or two days.

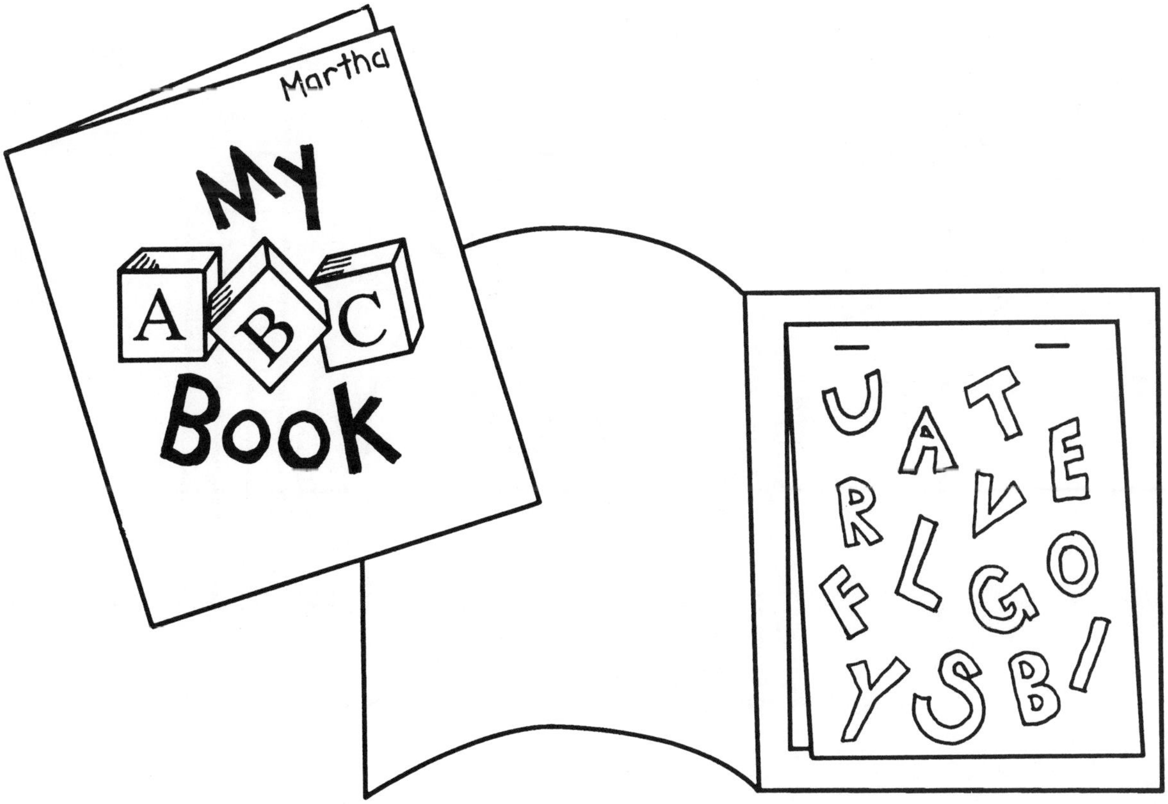

CLASS ALPHABET BOOK

Description of Booklet (Group Booklet)

This booklet uses two 12" x 18" sheets of colored construction paper or oak tag as a cover. The inner pages are 12" x 18" sheets of manila paper. The number of inner pages depends on the number of children in the room. The booklet is stapled together at one end.

Learning Skills Involved

Eye-hand coordination and auditory discrimination.

Method of Preparation

Each child selects one letter in the alphabet. It is printed on his/her 12" x 18" sheet of manila paper. Through individual instruction, the teacher emphasizes the sound of the letter to the child who then draws a picture of an object on the paper beginning with the letter.

The pictures are then arranged in alphabetical order by the teacher and stapled into the booklet. For a colorful effect, children can be given colored felt-tipped markers for drawing.

Preparation Time

Approximately one day.

MY ALPHABET BOOK

Description of Booklet

This booklet consists of a 12" x 18" cover of colored construction paper that is folded in half. Inner pages are worksheets that are prepared in advance by the teacher, inserted inside the cover and stapled along the folded side.

Learning Skills Involved

Eye-hand coordination, visual discrimination, and following directions.

Method of Preparation

Children are given a different worksheet each day to complete. When finished, these are collected by the teacher who saves them and staples them into booklets.

Covers may be decorated.

Preparation Time

One day per worksheet.

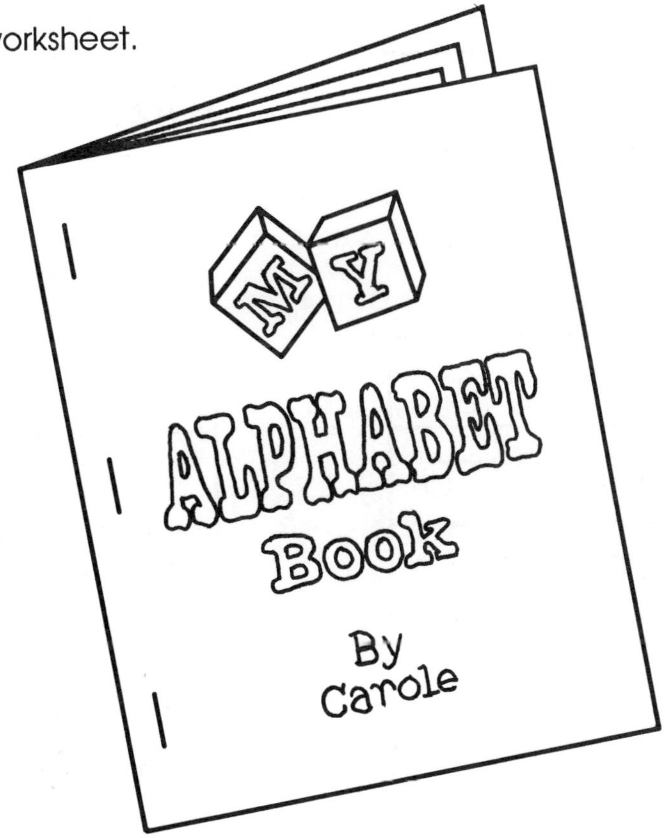

Suggestions for worksheets are given on the next page.

MY ALPHABET BOOK

Suggestions for teacher-made worksheets:

Copy the letters	
a	
b	
c	
d	
e	

Circle the same letters	
aba	rmv aba blt
bcd	bcd dcb bad
rus	sur sir rus
vip	vip viv piv
bad	dad bad mop

Circle the same letter	
r	lprsuvro
s	qstmwbi
t	tmtsyvw
u	laudamu
v	rtvuvvy

Circle all the "d"s

PICK A SUBJECT

Description of Booklet

These are novelty booklets constructed in advance by the teacher. Covers consist of colored construction paper with inner pages of typing paper.

Learning Skills Involved

Eye-hand coordination and creativity.

Method of Preparation

Children are given blank books in the shape of a subject such as "GHOSTS," "CARS," "BUTTERFLIES," "DOLLS," etc. A selection of various topics may be offered and children choose the one that interests them most. They may write a story or draw pictures to illustrate their booklet.

Preparation Time

One or two days.

WORDS AROUND THE ROOM

Description of Booklet

This booklet consists of a 12" x 18" cover of colored construction paper that is folded in half. Several lined inner pages are stapled inside the back cover.

Learning Skills Involved

Eye-hand coordination and visual discrimination.

Method of Preparation

Using felt marking pens and sentence strips or plain paper strips, the teacher prints labels for as many items in the room as possible and tapes them to the objects; i.e. door, desk, chair, paper, crayons, scissors, window, wall, light, plant, clock, chalk, chalkboard, coat rack, cubby, toy shelf, books, sink, etc.

Children are asked to walk around the room and copy as many words as possible in their booklets.

Preparation Time

One or two days.

DIFFERENT WAYS OF WRITING

Description of Booklet

This booklet consists of a 12" x 18" cover of colored construction paper that is folded in half. Several lined pages are stapled to the inner back cover.

Learning Skills Involved

Eye-hand coordination.

Method of Preparation

Books or papers containing prose or poetry are made available to students for copying. Each poem or paragraph that is copied must be written in either manuscript, cursive, D'Nealian or other type of script. Children may execute one page per day which the teacher collects, saves, and then staples into the booklet.

Covers may be decorated.

Preparation Time

Approximately three or four days.

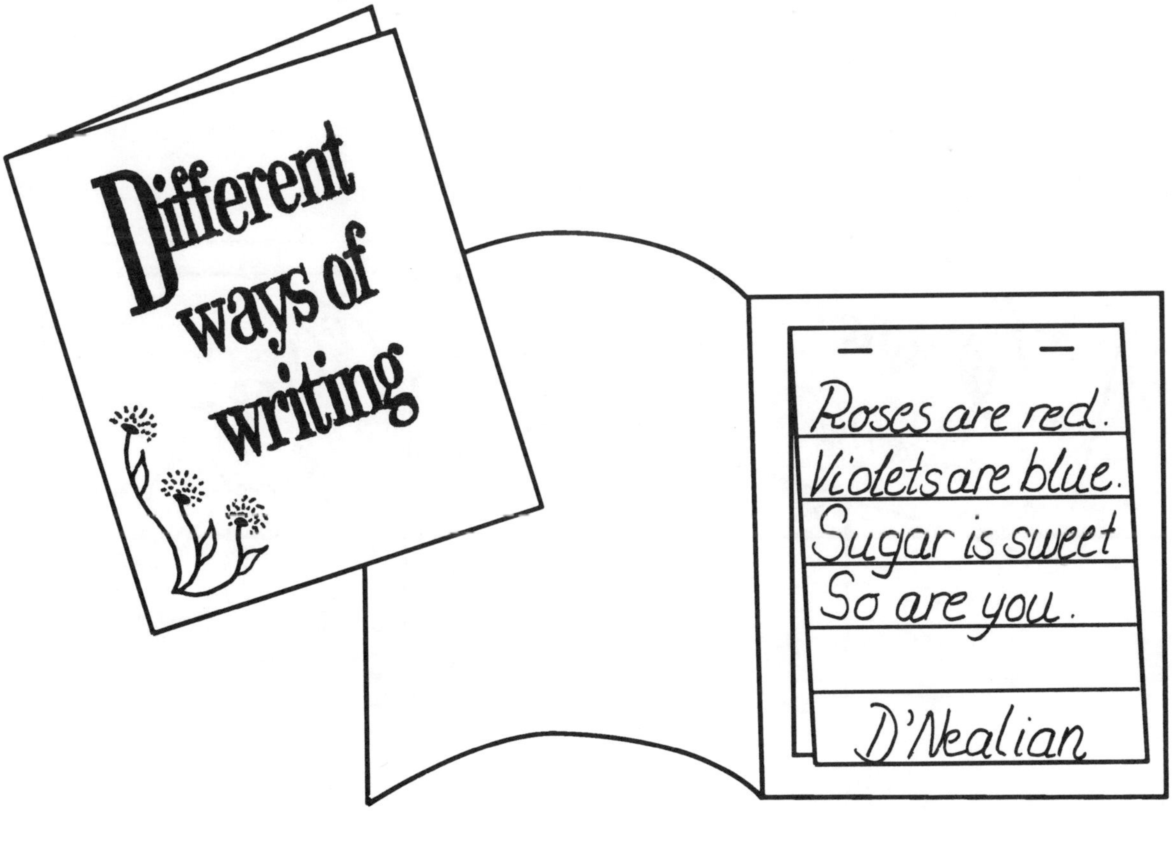

ILLUSTRATE A NURSERY RHYME

Description of Booklet

This booklet consists of a 12" x 18" cover of colored construction paper that is folded in half. Pages of typing paper are stapled to the inner back cover.

Learning Skills Involved

Listening, following directions, making decisions, and creativity.

Method of Preparation

The teacher reads several nursery rhymes to the children. They are then given blank booklets, crayons, and pencils and asked to illustrate one rhyme from beginning to the end. The teacher, or if possible the student, writes the appropriate line at the bottom of each page.

Preparation Time

One day.

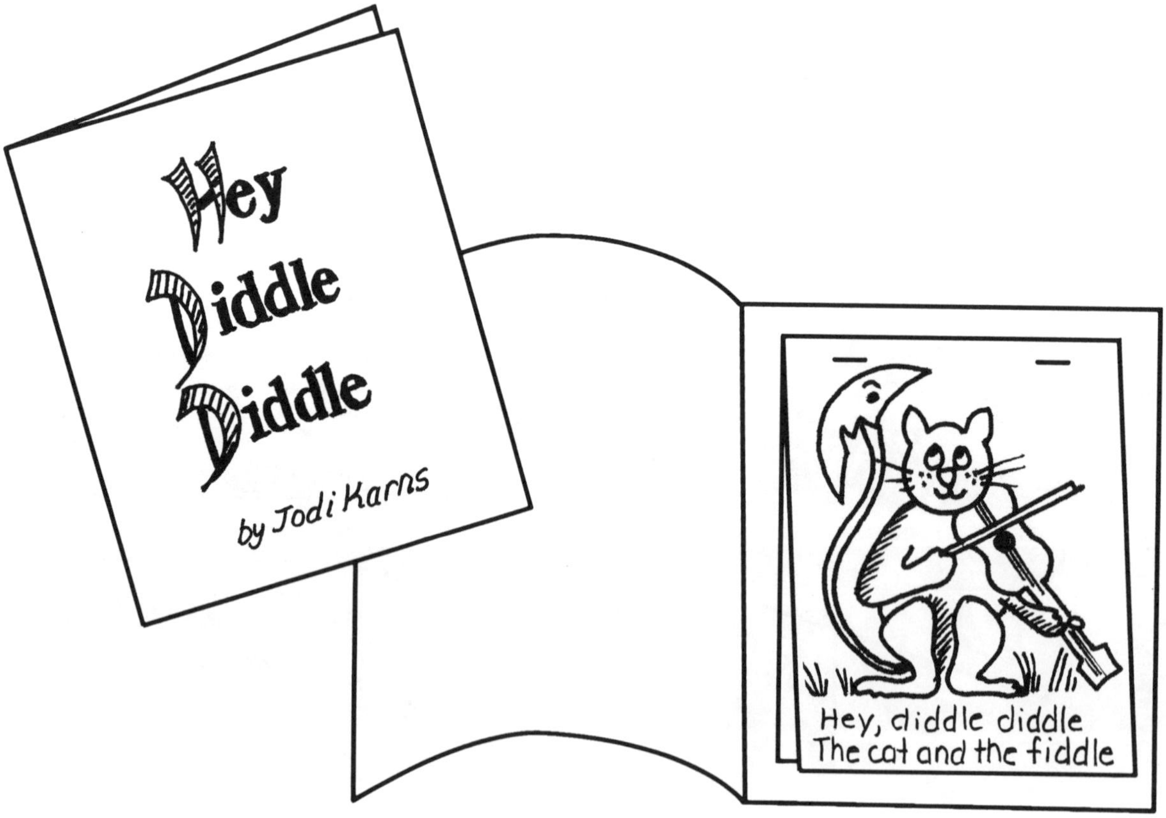

WORDS THAT RHYME

Description of the Booklet

This booklet consists of a 12" x 18" cover of colored construction paper that is folded in half.

Inner pages are worksheets that are prepared in advance by the teacher, inserted inside the cover and stapled along the folded edge.

Learning Skills Involved

Following directions, making decisions, and rhyming.

Method of Preparation

A key word is printed at the top of each worksheet by the teacher. One syllable words may be used such as: at, door, sing, rug, or pen, etc.

The child writes as many words as possible that rhyme with the key word on the page. As pages are completed, they are collected by the teacher and then stapled into a booklet.

Suggestion: Key words may be individualized for each child.

Preparation Time

One page per day.

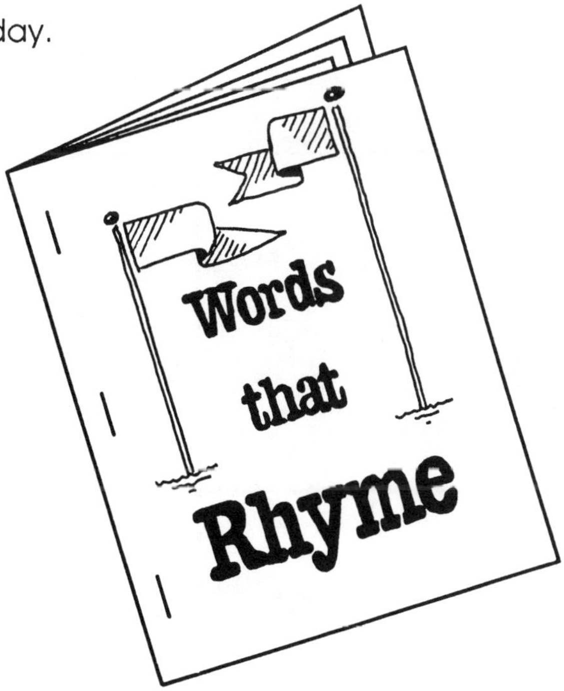

Examples are given on the next page.

WORDS THAT RHYME

Suggestions for teacher-made worksheets:

fat	
cat	scat
sat	
mat	
rat	
bat	
gnat	

pear	
bear	stair
fair	
tear	
wear	
lair	

star	
bar	far
car	jar
tar	

bell	
fell	well
dwell	tell
sell	swell

ANT, BEE AND CAT

Description of Booklet

This booklet consists of a 12" x 18" cover of colored construction paper that is folded in half. Three inner pages will be stapled inside the cover upon completion by students.

Learning Skills Involved

Following directions, observing, and identifying.

Method of Preparation

Children are given an inner page each day with either "ant," "bee," or "cat" as a keyword printed at the top. On each page the child lists as many words as possible containing the key word. For example, if the word is ant; the list might contain words such as plant, pleasant, anthem, frantic, slant, fantastic, etc. Dictionaries may be used.

As pages are completed, they are collected by the teacher who saves them and staples them into a booklet.

Preparation Time

Three days.

MY WISH BOOK

Description of Booklet

This booklet consists of a 12" x 18" cover of construction paper that is folded in half. As inner pages are completed by students, they will be stapled inside the cover.

Learning Skills Involved

Following directions and creativity.

Method of Preparation

A general class discussion is recommended to address the wishes and things children might want. They are then required to either draw or describe in a paragraph a different wish on each page. As pages are completed, they are collected by the teacher who saves them and staples them into the booklet.

Preparation Time

From two to three days.

FLOWERS

Description of Booklet (Group Booklet)

This booklet consists of two 12" x 18" cover pages of oak tag and inner pages of 12" x 18" manila paper. As students finish their pages, the teacher staples them inside the cover at one end.

Learning Skills Involved

Eye-hand coordination, identification, and visual discrimination.

Method of Preparation

Each child receives a sheet of manila paper on which he/she may draw flowers. Crayons and/or colored felt markers are furnished along with pencils. Creativity is encouraged. Children may wish to copy flowers from seed catalogues or magazines. Flowers may be labeled.

Preparation Time

One day. The booklet should be placed on the bookshelf in the classroom.

SEQUENCE

Description of Booklet

This booklet consists of two 12" x 18" cover pages of colored construction paper and several 12" x 18" sheets of manila paper as inner pages. The booklet is stapled together at one end.

Learning Skills Involved

Following directions, arranging in order, and collecting.

Method of Preparation

Children are asked to save and bring to school several comic strips from the newspaper. The strips are cut up and each is placed in an envelope.

The child, working on a booklet, selects one envelope and pastes it in sequential order on an inner page. When completed, the child selects another envelope.

It is suggested that the child paste at least three different comic strips in order for a booklet.

Preparation Time

One day.

Suggestion: The teacher should begin saving simple comic strips before this booklet is offered to the class. These will supplement what the children bring.

MY BOOK OF RULES

Description of Booklet

This booklet consists of a 9" x 12" cover page that is folded in half. Several lined pages are stapled to the inner back cover.

Learning Skills Involved

Following directions, observing, collecting, identifying, and expressing emotions.

Method of Preparation

Blank booklets are constructed in advance by the teacher.

A class discussion is held and classroom rules suggested. Children are then asked to list what they think would be suitable rules for the class.

Preparation Time

One day.

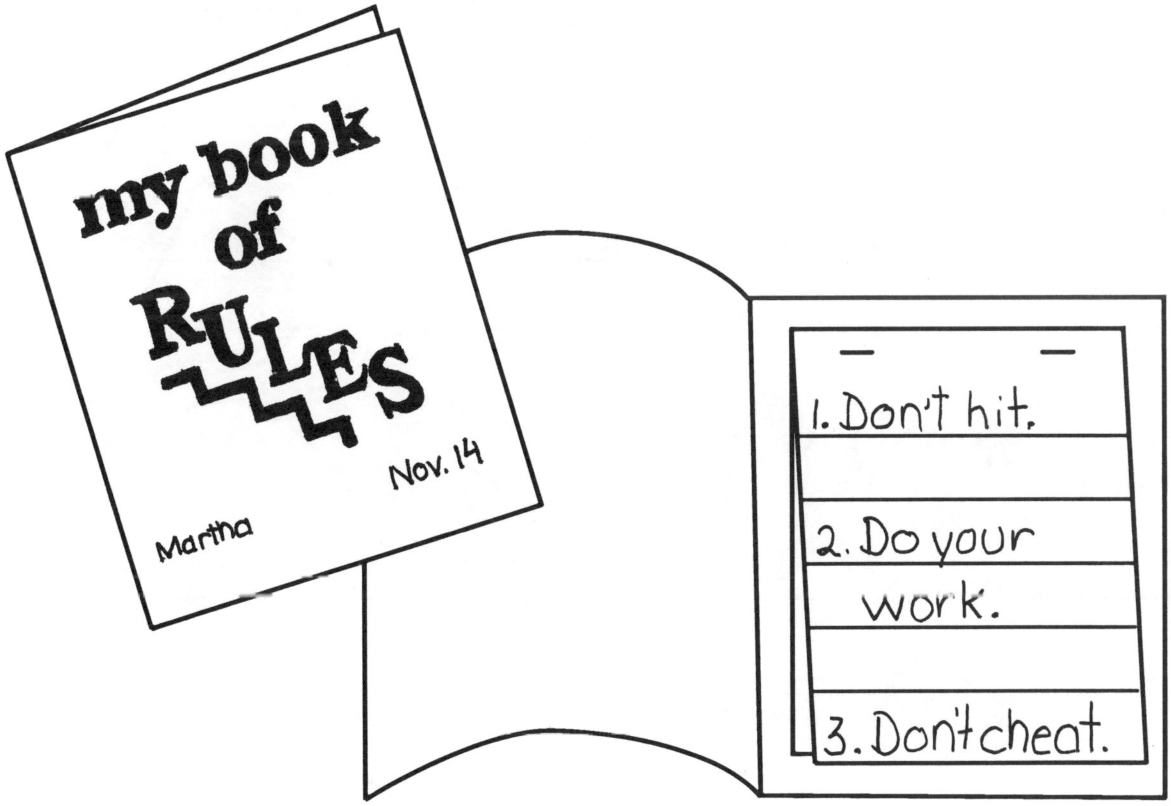

OUR TRIP TO THE ZOO

Description of Booklet (Group Booklet)

This booklet consists of two 12" x 18" cover pages of oak tag and 12" x 18" sheets of manila paper as inner pages. When assembled, the booklet is braded together at one end.

Learning Skills Involved

Visual and auditory discrimination, memory, observation skills, following directions, and summarizing.

Method of Preparation

Following a class field trip, each child is given a sheet of 12" x 18" manila paper and asked to draw one aspect of the trip. Children can be assigned specific occurrences such as the bus trip to the zoo, a particular animal, or they can draw what impressed them most. Crayons and felt tipped markers enhance drawings.

When completed, the teacher arranges the papers and brads them into a booklet that can be kept in the classroom on the bookshelf for personal reading. Either the teacher or the children should write an explanation of each picture on every page.

Preparation Time

One day.

VERBS

Description of Booklet

This booklet consists of a 12" x 18" cover of colored construction paper that is folded in half. Several pages of typing paper will be stapled inside the cover after they have been completed by students.

Learning Skills Involved

Eye-hand coordination, following directions, and focusing in on the subject.

Method of Preparation

The teacher prints a different "action-type" verb at the bottom of many sheets of typing paper. Children are furnished with scissors, paste, and old magazines.

Each child selects a sheet and searches through magazines for a picture that might illustrate the verb. The picture is cut out and pasted on the page. After several pages are completed, the teacher staples them into a booklet.

Preparation Time

One or two days.

MY BOOK OF LONG VOWELS

Description of Booklet

This booklet consists of a cover made of two sheets of 9" x 12" colored construction paper. Several pages of typing paper will be stapled inside the cover after they have been completed by students.

Learning Skills Involved

Auditory discrimination

Method of Preparation

Children print a vowel at the top of each page. Under the vowel, they list as many words as possible that use the long sound of the vowel. The teacher collects and saves completed papers and then staples them into a booklet.

This same type of booklet can be adapted for short vowels.

Preparation Time

Five days – one for each vowel.

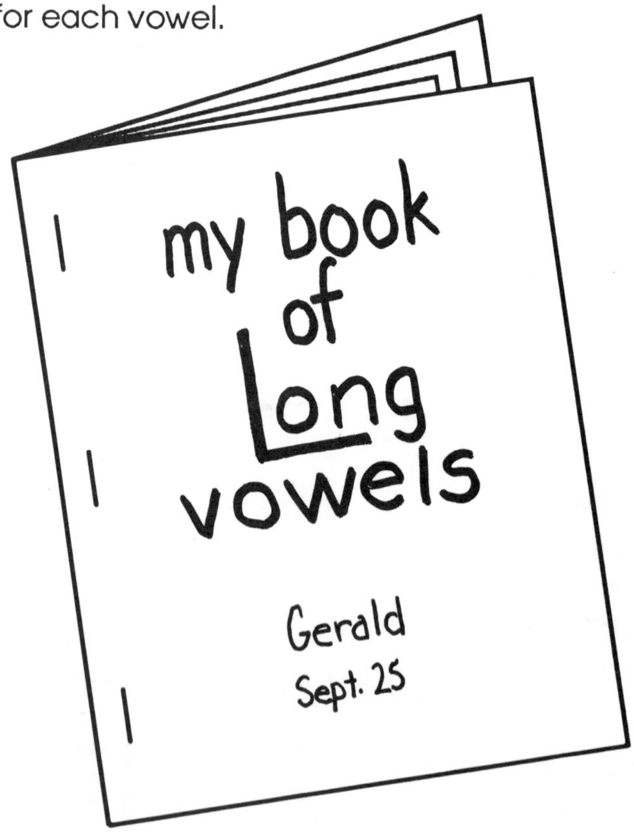

Examples of this type of booklet are shown on the following page.

Examples for

MY BOOK OF LONG VOWELS

a

bake lake
fable
ate ape
mate
race take
rate
cane plane

e

easel me
fee we
feet eat
sweet seat
wheat
weed easy
repeat

i

kite fight
sight ice
light idea
might
fine wine
dine time
sign iron

o

ocean oats
poach
pose rose
scold fold
roll rogue
vote no

GET WELL

Description of Booklet (Group Booklet)

This booklet is written by the children to be sent home to a sick schoolmate. The cover consists of a 12" x 18" sheet of colored construction paper that is folded in half. Typing paper will be used as inner pages after they have been completed by students.

Learning Skills Involved

Eye-hand coordination and creativity.

Method of Preparation

Each child is given a sheet of typing paper on which he or she writes a message or poem or draws a picture for the sick child.

As they are finished, the teacher collects the papers and when all are turned in they are stapled into the booklet.

Preparation Time

One day.

OPEN IN THE YEAR 2010

Description of Booklet

This booklet consists of a 12" x 18" cover of oak tag that is folded in half. The inner pages are 8 1/2" x 11" sheets of typing paper or lined paper.

Learning Skills Involved

Decision making, observation, hypothesizing, and creativity.

Methods of Preparation (2)

Children are given paper, pencils and crayons or felt markers. The instructor decides which of the following methods to use:

1. Children discuss what they think the year 2010 will be like and brainstorm their booklets. For example, they may describe future houses, automobiles, airplanes, clothing, telephones, television, etc.

2. Children discuss meaningful persons, places or things in their lives today. Ideas such as favorite food, sports figures, TV programs, cars, music, hobbies, friends, etc.

Covers are decorated and each booklet sealed with tape. When opened in the year 2010, they can recall the past or see if their predictions were accurate.

Preparation Time

Four or five days.

CRAZY, MIXED-UP WORDS

Description of Booklet

This booklet consists of a 12" x 18" cover of colored construction paper that is folded in half. Worksheets are stapled to the inner back cover.

Learning Skills Involved

Observation, identification, and rearranging.

Method of Preparation

The teacher prepares worksheets, each consisting of lists of words containing mixed-up letters. Each worksheet pertains to a specific category. For example, if the category is a SCHOOL, the words might be: ksde, (desk); nipcel, (pencil); clakh, (chalk). Children must fill in the correctly spelled word.

Categories might include ANIMALS, CIRCUS, TREES, FLOWERS, SPORTS, FOOD, SUMMER, WINTER, etc.

After filling in several worksheets, the teacher staples the pages inside the cover. (Covers may be decorated.)

Preparation Time

One worksheet per day per child. Four or five days would be needed to finish the project.

A sample of a worksheet for CRAZY, MIXED-UP WORDS is given on the following page.

CRAZY, MIXED-UP WORDS

Worksheet

____

(Category)

Name ______________________________

MY STORY

Description of Booklet

This booklet consists of a 12" x 18" cover of colored construction paper that is folded in half. Typing paper will be used as inner pages.

Learning Skills Involved

Oral language development and creativity.

Method of Preparation

Children take turns dictating their stories to the teacher. (A good place to engage one child at a time in conversation is on the playground.) Teachers should offer suggestions when needed and try to limit the contents. Older children may write their own stories.

Illustrations should be encouraged.

Preparation Time

Two to five days.

An example of this kind of booklet is given on the following page.

Examples for

MY STORY

1

2

3

4

Older children may write and illustrate their own books. Teacher assistance should be given when the child requests it.

FINISH THE STORY

Description of Booklet

This booklet consists of a 12" x 18" cover of colored construction paper that is folded in half. Several inner pages are inserted and stapled in the center. For decoration, two holes may be punched along the fold and yarn inserted by the student.

Learning Skills Involved

Imagination and creativity.

Method of Preparation

The teacher reads approximately one-fourth of a story (either a new story or a fairy tale) to the class. Children are then given blank booklets, yarn, hole puncher, crayons, markers and/or pencils. They illustrate and finish the story with their own unique endings.

Preparation Time

Two or three days.

BOOK REPORTS

Description of Booklet

This booklet consists of a 12" x 18" cover of colored construction paper that is folded in half. Typing paper is used for the inner pages that are stapled along the fold.

Learning Skills Involved

Identifying and summarizing.

Method of Preparation

Blank booklets are provided by the teacher.

After reading a book, the child takes a blank booklet and writes and illustrates a report. Crayons, pencils and markers are offered.

Children can be encouraged to build a collection of book report booklets.

Preparation Time

One or two days.

MY BOOK OF WORDS FOR EACH SEASON

Description of Booklet

This booklet consists of a 12" x 18" cover of manila paper that is folded in half. Four inner pages are stapled inside the back cover.

Learning Skills Involved

Eye-hand coordination, observation and imagination.

Method of Preparation

Each page is titled either "summer," "autumn," "winter" or "spring." Some type of illustration is drawn and colored on each page and a list of descriptive words is written on the same page.

Preparation Time

Four days.

Ideas for this booklet are given on the following page.

Ideas for

MY BOOK OF WORDS FOR EACH SEASON

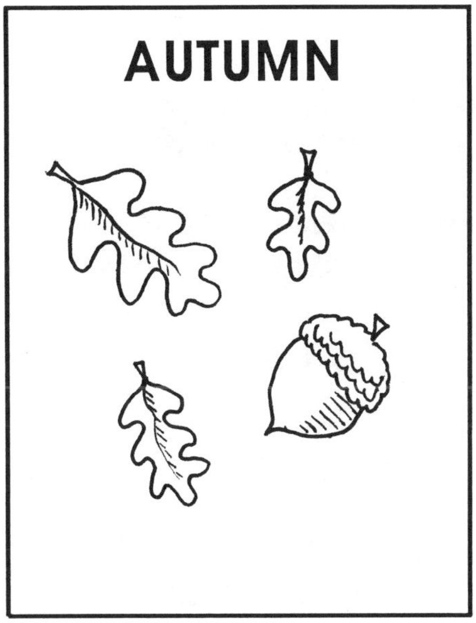

MY OCTOBER BOOK

(or any other month)

Description of Booklet

The booklet consists of a 12" x 18" cover of construction paper or oak tag that is folded in half. Several inner pages of worksheets will be stapled inside the cover after students have completed them.

Learning Skills Involved

Following directions, summarizing, making decisions, and classifying.

Method of Preparation

The teacher prepares several worksheets to be filled in by students. They may draw pictures and/or write descriptive sentences about the suggested topic at the top of each page.

Papers are collected by the teacher. When all work is completed, the pages are stapled into the booklets.

Preparation time

Three to five days.

Ideas for worksheets are given on the following page.

Ideas for

MY OCTOBER BOOK

Weather

Birthdays and Holidays

Kind of Clothing

Sports

MY BOOK OF POEMS

Description of Booklet

This booklet consists of a 12" x 18" cover of colored construction paper that is folded in half. Several lined pages are stapled to the inner back cover.

Learning Skills Involved

Eye-hand coordination and making decisions.

Method of Preparation

Booklets are constructed in advance by the teacher.

Several poetry books or single sheets of poems are made available to students. They select the poems they like best and copy them in their booklets. Illustrations should be encouraged.

Preparation Time

Three to five days. Children copy one poem a day. Class discussions may be held so children can explain why they selected a certain poem.

FINDING VOWELS

Description of Booklet

This booklet consists of a 12" x 18" cover of colored construction paper that is folded in half. Five inner pages are cut from a magazine such as *Time, Newsweek* or *U.S. News and World Report* are inserted in the cover and stapled along the fold.

Learning Skills Involved

Identification and observation.

Method of Preparation

Booklets are constructed in advance by the teacher. Five magazine pages are inserted and stapled inside the cover. Using a brightly colored felt marker, the teacher prints the name of a different vowel at the top of each page – using one side only.

Using highlighter pens, students color all the vowels on the page as indicated by the teacher's vowel at the top.

Preparation Time

Five days – one for each of the five vowels.

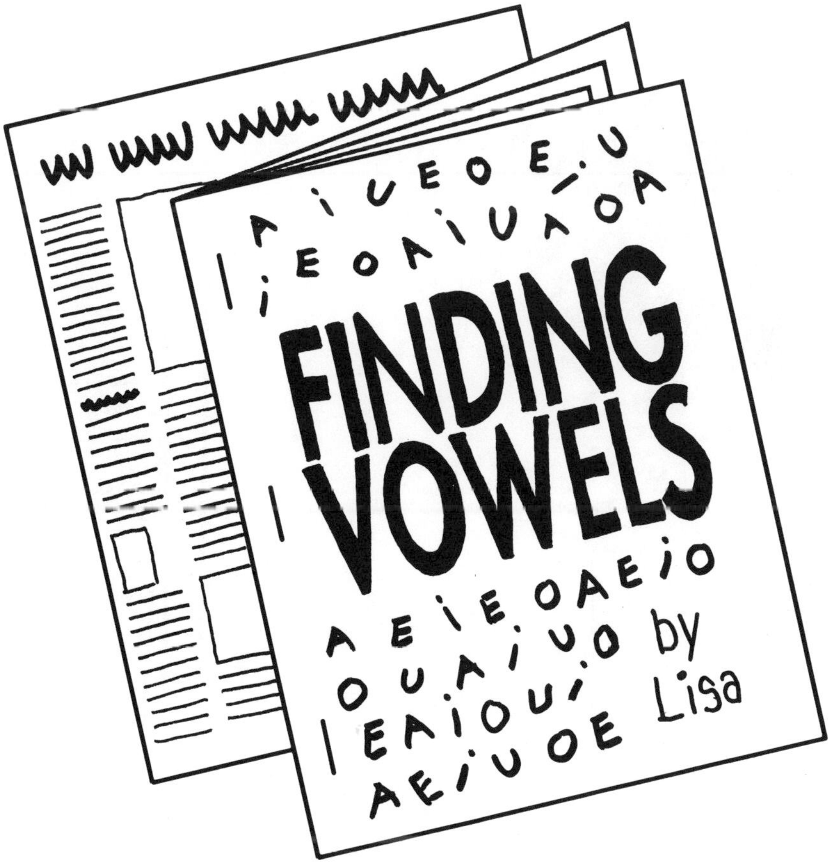

UP WITH VOWELS

Description of Booklet

This booklet consists of a 12" x 18" cover of colored construction paper that is folded in half. Several lined inner pages will be stapled inside the cover after students have completed them.

Learning Skills Involved

Eye-hand coordination, memory, and visual discrimination.

Method of preparation

Word lists that supplement curriculum subjects are prepared by the teacher. Lined paper and pencils are provided.

The child selects a word list, pastes it on the inside cover of his/her booklet and copies the words on lined paper. Vowels are written a step up from the rest of the letters in each word.

Preparation Time

Approximately three days. It is suggested that one word list a day per child be used.

MY "O" BOOK

(or any other vowel)

Description of Booklet

This booklet consists of a 12" x 18" cover of colored construction paper that is folded in half. Manila inner pages are stapled inside the cover. Booklets are made in the shape of the vowels they represent.

Learning Skills Involved

Eye-hand coordination and memory.

Method of Preparation

Booklets are constructed in advance by the teacher in the shape of vowels. Children select a vowel and draw and color items beginning with the same vowel on each page.

Variation #1: Old magazines are provided and children cut out pictures of items beginning with their vowel and paste them on the pages.

Variation #2: Newspaper columns are offered the children. They underline all the vowels that they have selected and paste them on their pages.

Preparation Time

One to five days.

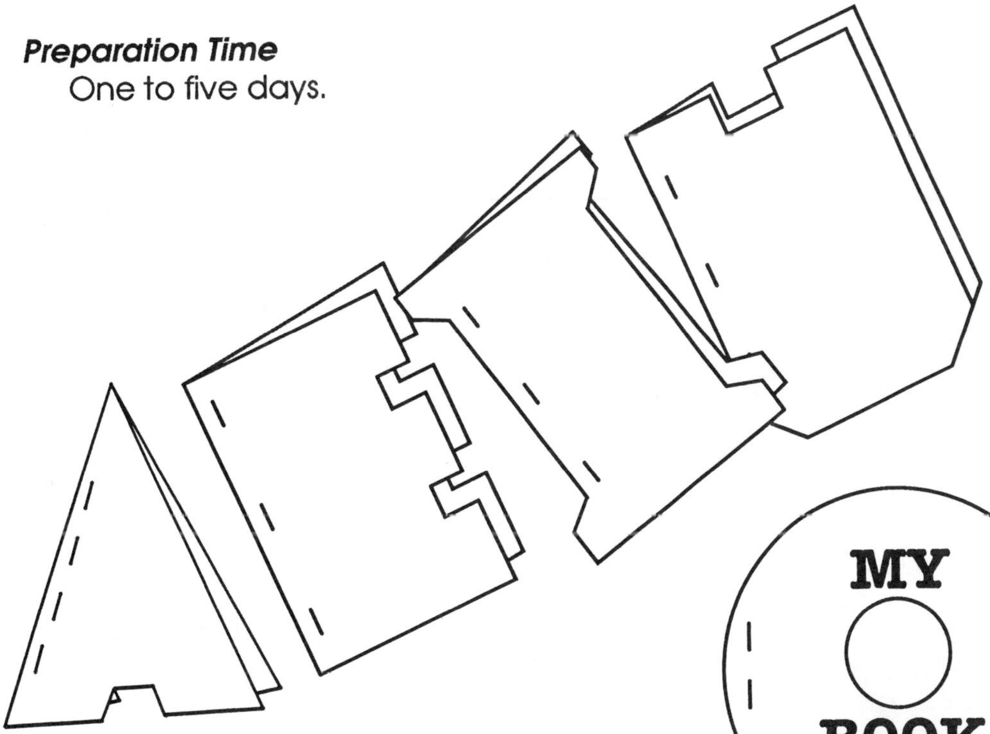

Chapter 3

MATHEMATICS

The activities in this chapter are planned to enhance the following skills:

- copying and identifying numerals
- one-to-one correspondence with numbers
- identifying shapes
- making a calendar
- telling time
- counting
- identifying fractions
- measuring

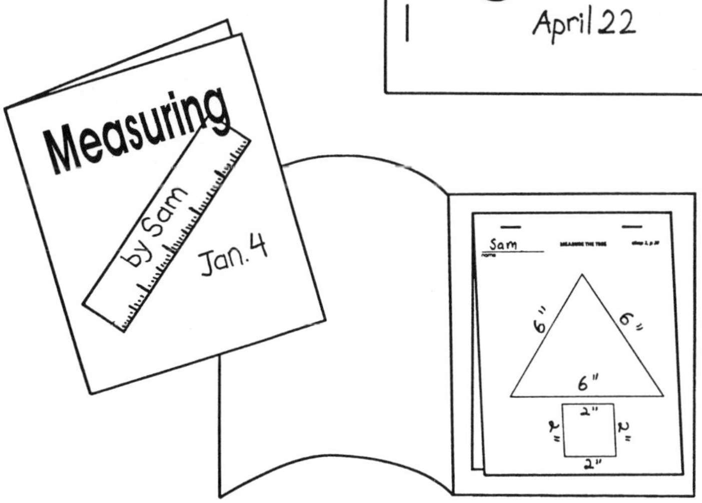

MY BOOK OF NUMBERS (1-5)

Description of Booklet

This booklet consists of a 12" x 18" cover of colored construction paper that is folded in half. Five pages of typing paper are stapled inside the fold.

Learning Skills Involved

Eye-hand coordination and thinking.

Method of Preparation

Booklets are constructed in advance by the teacher. At the top of each page, a numeral – either 1, 2, 3, 4, or 5 – is printed. Students draw the number of items that are indicated by the numeral. They may draw objects such as toys, cars, shapes, etc; or they may draw animals or people. All creative efforts should be encouraged.

Preparation Time

One or two days.

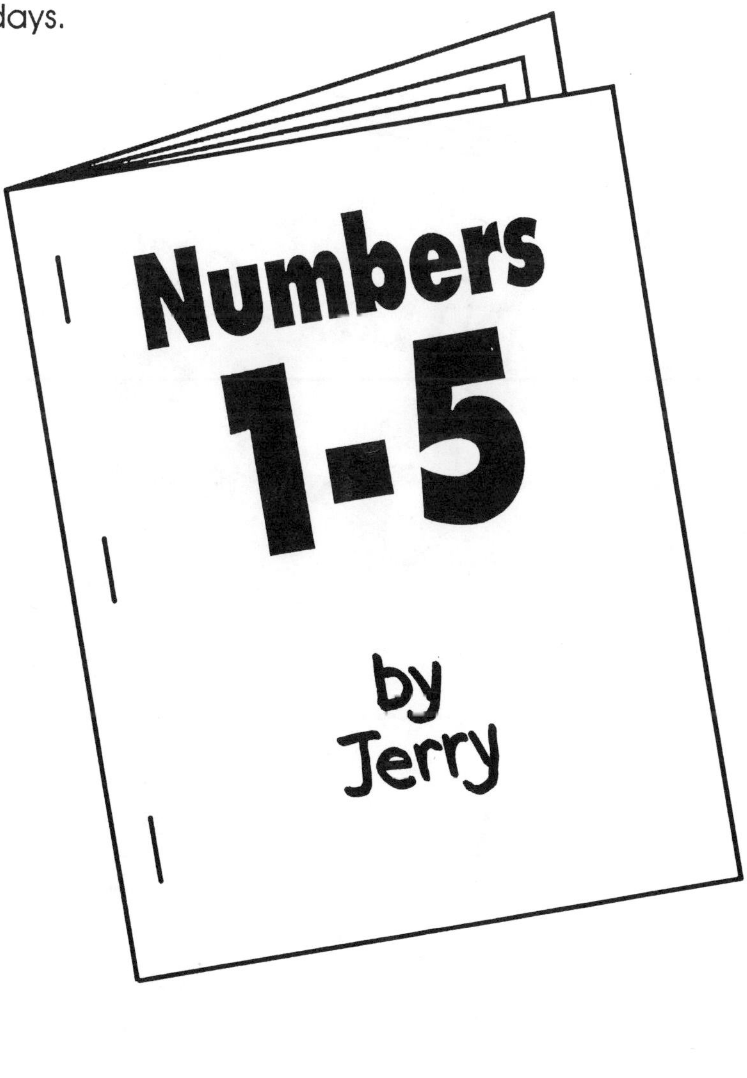

MY BOOK OF NUMBERS (6-10)

Description of Booklet

This booklet consists of a 12" x 18" cover of colored construction paper that is folded in half. Five numbered inner pages of typing paper are stapled inside the fold.

Learning Skills Involved

Eye-hand coordination and thinking.

Method of Preparation

Booklets are constructed in advance by the teacher. At the top of each page, a numeral – either 6, 7, 8, 9, or 10 – is printed. Students draw the number of items indicated by the printed numeral.

Preparation Time

One or two days.

COPY THE NUMERALS

Description of Booklet

This booklet consists of a 12" x 18" cover of colored construction paper that is folded in half. Inner pages are worksheets that have been prepared by the teacher.

Learning Skills Involved

Eye-hand coordination.

Method of preparation

Four different worksheets are prepared in advance by the teacher. The first contains numerals 1-5; the second contains numerals 6-10; the third contains numerals 11-15, and the fourth contains numerals 16-20. Students copy the given numerals across each line.

Preparation Time

Four days.

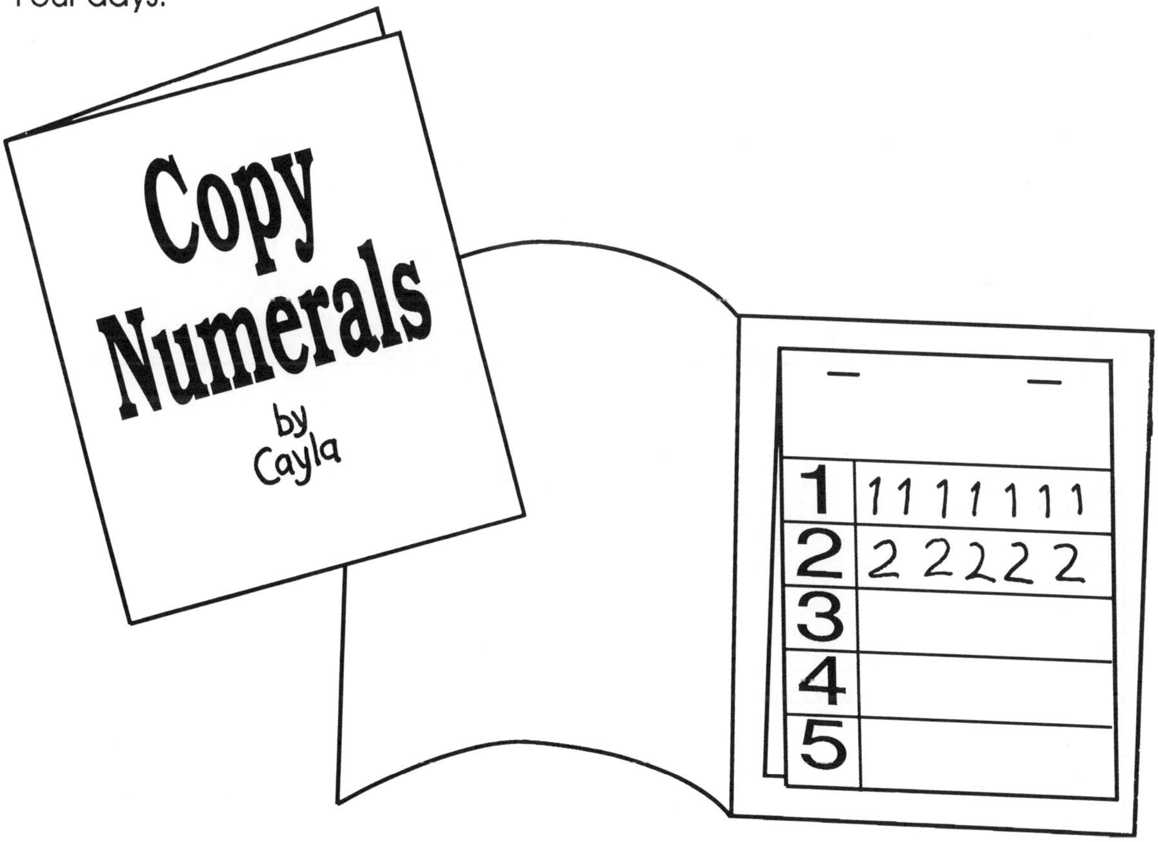

A sample worksheet is given on the following page.

COPY THE NUMERALS

Worksheet

Name ______________________________

HOUSE NUMBERS

Description of Booklet

This booklet consists of a 12" x 18" cover of colored construction paper that is folded in half. Several lined pages are stapled to the inner back cover.

Learning Skills Involved

Listening and following directions.

Method of Preparation

Booklets are constructed in advance by the teacher. Children ask their school friends to write their first name and street numbers (or apartment numbers) on a line in the book. They should solicit the entire class.

Preparation Time

One or two days.

NUMBER MATCHING

Description of Booklet

Each booklet consists of a 9" x 12" cover of colored construction paper that is folded in half. Three folded sheets of typing paper are inserted and stapled inside the cover.

Learning Skills Involved

Counting and one-to-one correspondence.

Method of Preparation

Booklets are prepared in advance by the teacher. Each page is numbered from 1-12, using both front and back. Children are provided with self-stick gummed circles. They attach the proper number of circles on each numbered page.

Preparation Time

Some students finish in one day. Others may take from two to three days.

STAMP THE NUMBERS

Description of Booklet

This booklet consists of a 12" x 18" cover of colored construction paper that is folded in half. Three worksheets will be stapled inside the cover after students have completed them.

Learning Skills Involved

Eye-hand coordination and counting.

Method of Preparation

Worksheets are prepared in advance by the teacher. Using the sample worksheet given on the following page, the teacher writes different numbers in each of the empty stamps.

Children are provided with small stamps and stamp pads. They stamp the given number in each space.

One worksheet is offered per day for three days. When completed, the teacher collects the worksheets and staples them into each student's booklet.

Preparation Time

Three days for younger students. Older students may take less time.

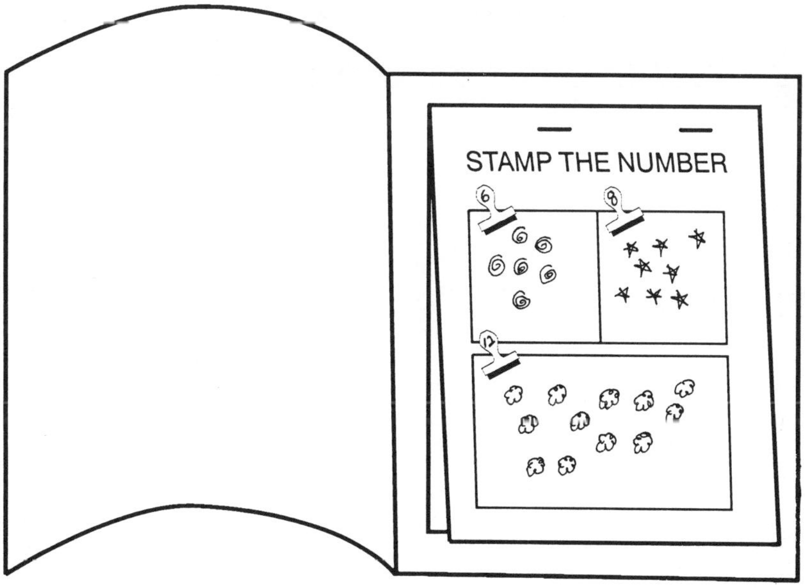

A sample worksheet is given on the following page. The bigger space is provided for larger numbers.

STAMP THE NUMBER

In Each Shape

Worksheet

Name ______________________________

FINGER COUNTING

Description of Booklet

This booklet consists of a 12" x 18" cover of colored construction paper that is folded in half. Five inner pages of typing paper are stapled to the inside back cover.

Learning Skills Involved

Eye-hand coordination and visual discrimination.

Method of Preparation

Booklets are prepared in advance by the teacher. Each day, children are asked to trace a hand with either one, two, three, four, or five fingers showing. Crayons and/or felt-tipped markers are placed on the work table. The number corresponding to the fingers is written by the child on each page.

Preparation Time

Five days. (A page a day.)

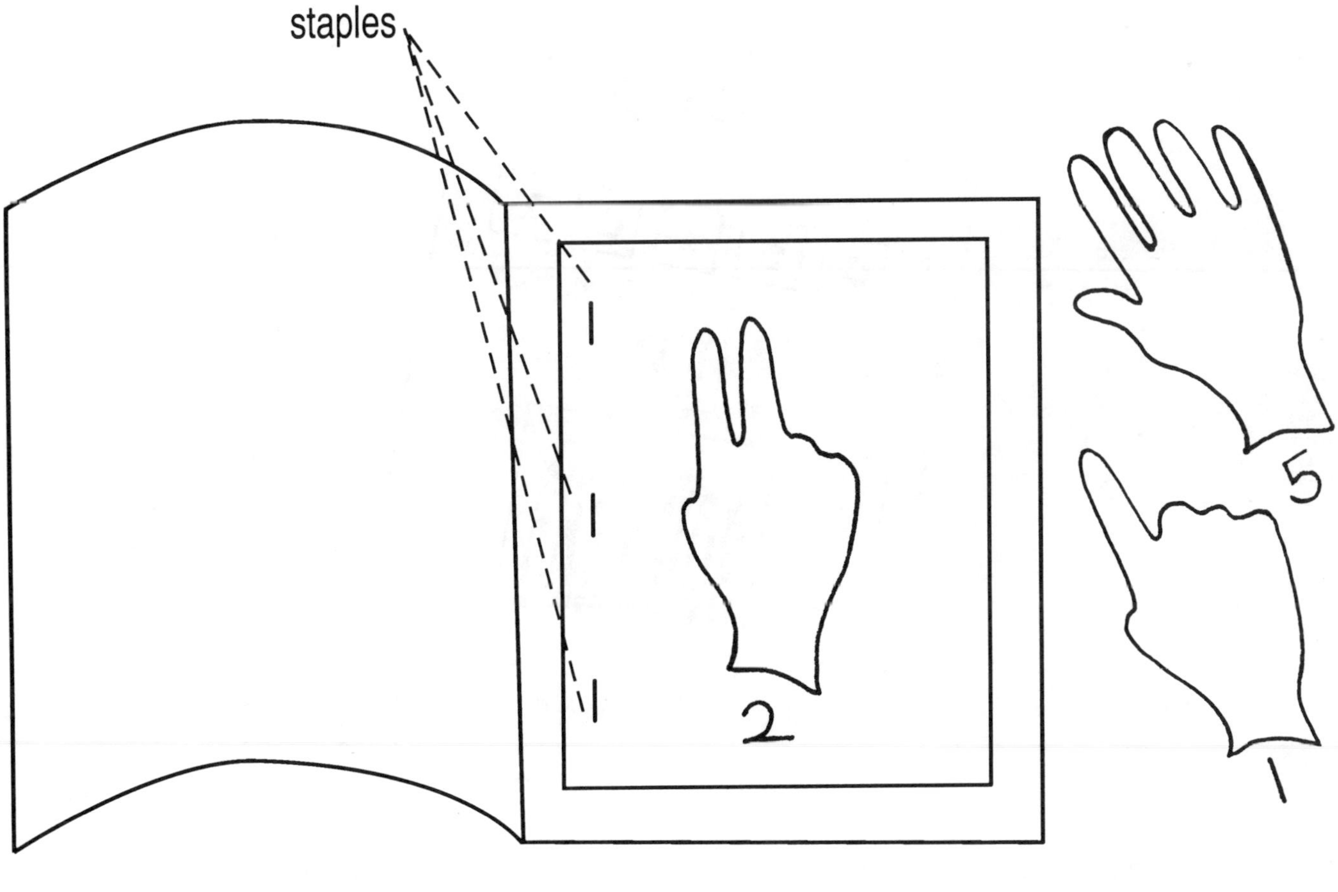

COLOR THE SQUARES #1

Description of Booklet

This booklet consists of a 12" x 18" cover of colored construction paper that is folded in half. Several worksheets will be stapled inside the cover after students have completed them.

Learning Skills Involved

Eye-hand coordination and problem solving.

Method of Preparation

Worksheets containing a grid with numbers written down the left-hand column are prepared in advance by the teacher. Students then color in the number of squares next to the indicated number. When completed, the teacher collects the worksheets and staples them into each student's booklet.

Preparation Time

Three to five days, depending how many worksheets will be offered. Each day's worksheet has numbers placed in different positions.

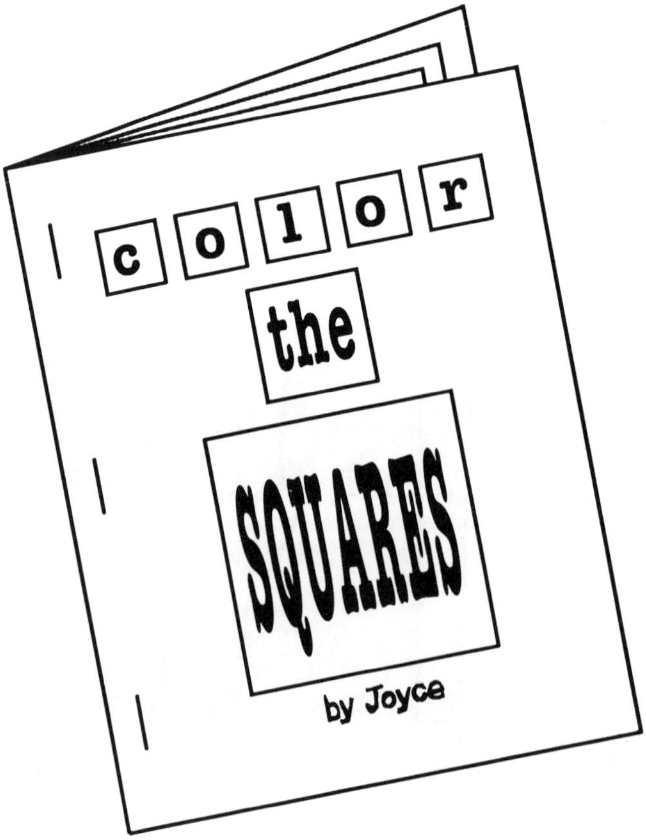

Examples of worksheets and a sample grid are given on the following pages.

Ideas for

COLOR THE SQUARES

name

2										
6										
1										
[illegible]										
10										
0										
4										
8										
5										
10										
4										
8										

name

10										
4										
3										
8										
9										
7										
6										
2										
0										
6										
9										
8										

name

3										
8										
6										
4										
1										
2										
9										
8										
2										
10										
6										
4										

name

4										
10										
5										
8										
7										
6										
2										
5										
3										
10										
7										
2										

Grid for

COLOR THE SQUARES

Name ______________________________

number

COLOR THE SQUARES #2

Description of Booklet

This booklet consists of a 12" x 18" cover of colored construction paper that is folded in half. Inner pages are worksheets to be inserted as they are completed.

A key list of numerals, each representing a different color, is constructed to accompany this activity.

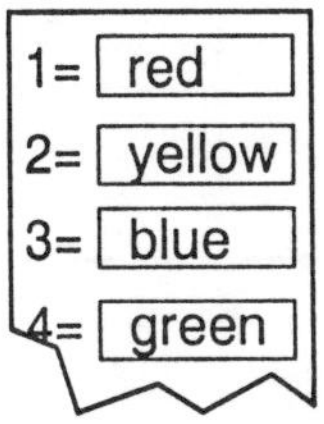

Learning Skills Involved

Observation and eye-hand coordination.

Method of Preparation

Booklet covers are prepared in advance along with worksheets and key lists. One or two key lists are placed prominently on the work table along with crayons.

Teachers prepare worksheets by randomly writing numerals in each of the grids on the worksheets. Numerals may be repeated or range as high as the number of spaces on the grid. Students then color each numbered square according to the color indicated on the key list.

The numerals on each day's worksheets may be rearranged.

Preparation Time

Younger children may spend more than one day on a worksheet. Older students can finish a worksheet each day. Depending on the number of worksheets offered, this activity could last from three to five days.

A grid is given on the following page.

COLOR THE SQUARES

MATCHING

Description of Booklet

This booklet consists of a 12" x 18" cover of colored construction paper that if folded in half. Teacher-made worksheets are stapled to the inner back cover.

Learning Skills Involved

Eye-hand coordination and observation.

Method of Preparation

Worksheets are prepared in advance by the teacher. Objects or animals are shown. Missing numbers or parts are to be filled in on each worksheet.

Preparation Time

Four days.

Four sample worksheets are given on the following pages.

FISH WORKSHEET

Name ______________________________

Cut out tails and paste on the matching fish.

LADY BUG WORKSHEET

Name ______________________________

Cut out heads and paste on matching bodies.

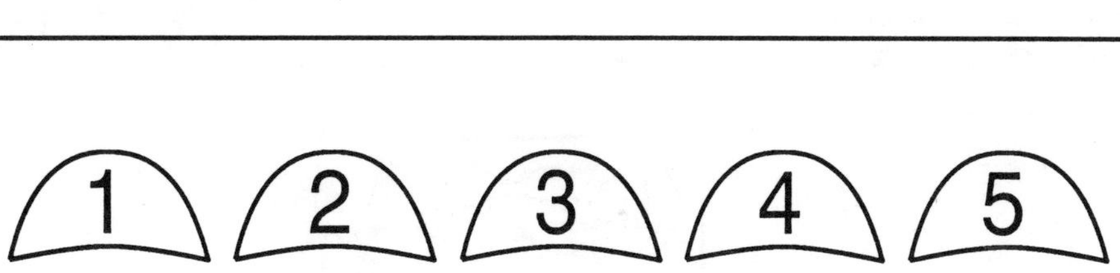

BIRTHDAY CAKE WORKSHEET

Name ______________________________

To each cake, add the number of candles as indicated in the oval.

HOT AIR BALLOONS WORKSHEET

Name ___________________________________

Count the number of stars in each hot air balloon and write in the matching number in each basket.

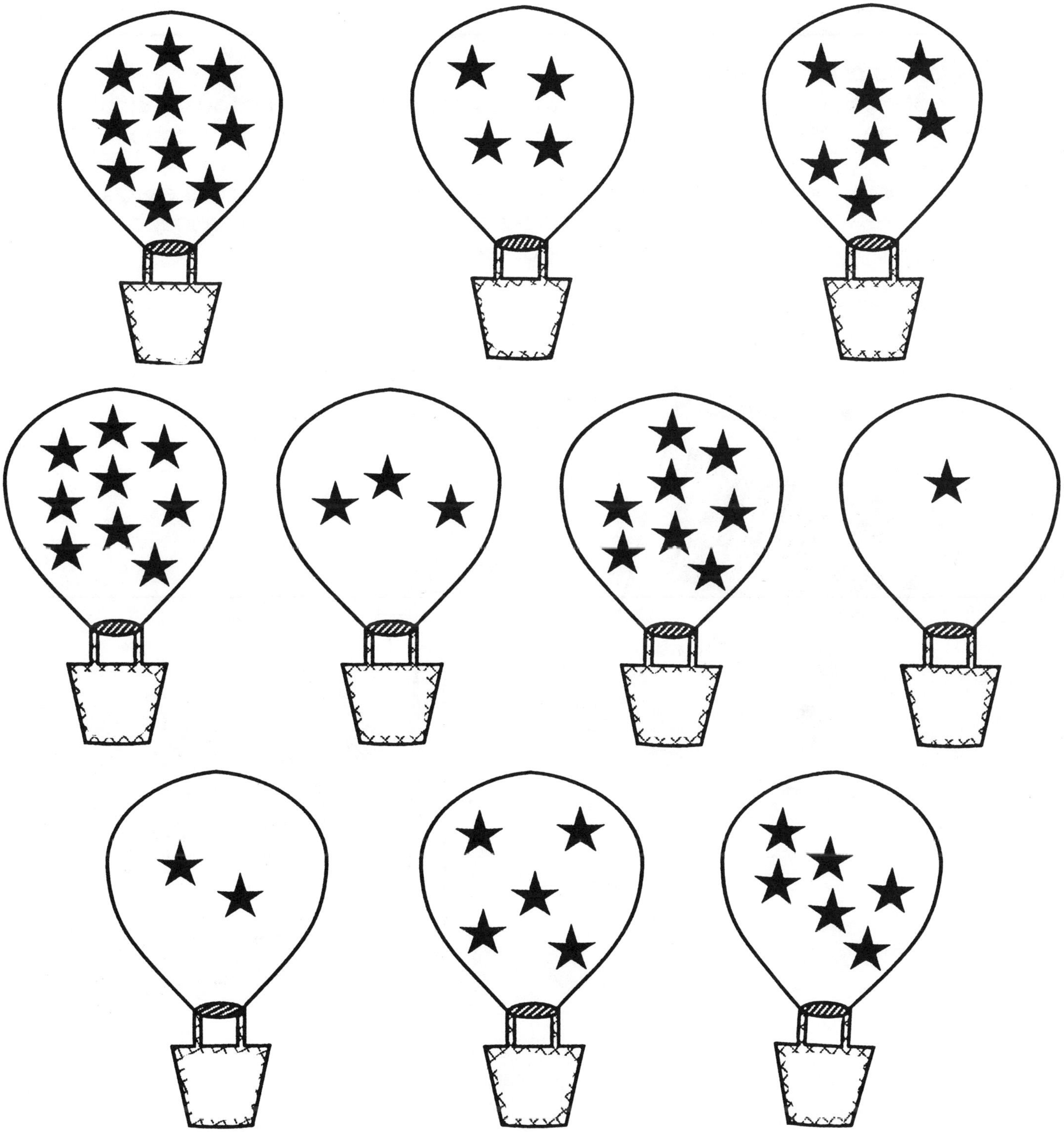

MY BOOK OF SHAPES

Description of Booklet

This booklet consists of a 12" x 18" cover of colored construction paper that is folded in half. Several pages are stapled to the inner back cover.

Learning Skills Involved

Identification.

Method of Preparation

Booklets are constructed in advance by the teacher. At the top of each page the teacher draws a shape or writes the name of the shape or both.

Many various-sized shapes are placed randomly in the center of the work table. Children search through the pile and select the matching shape or shapes and paste them on their papers.

Preparation Time

One or two days.

DRAW A SHAPE

Description of Booklet

This booklet consists of a 12" x 18" cover of colored construction paper that is folded in half. Several worksheets are stapled to the inner back cover when they have been completed by the students.

Learning Skills Involved

Eye-hand coordination.

Method of Preparation

Shape worksheets are prepared in advance by the teacher. Each worksheet contains a single small shape in the center of the page. Children are asked to draw around the shape. Each worksheet focuses on a different shape: circle, rectangle, triangle, square and oval.

When a worksheet is completed, the teacher collects it and after five days all worksheets will be stapled inside the cover.

Preparation Time

Five days.

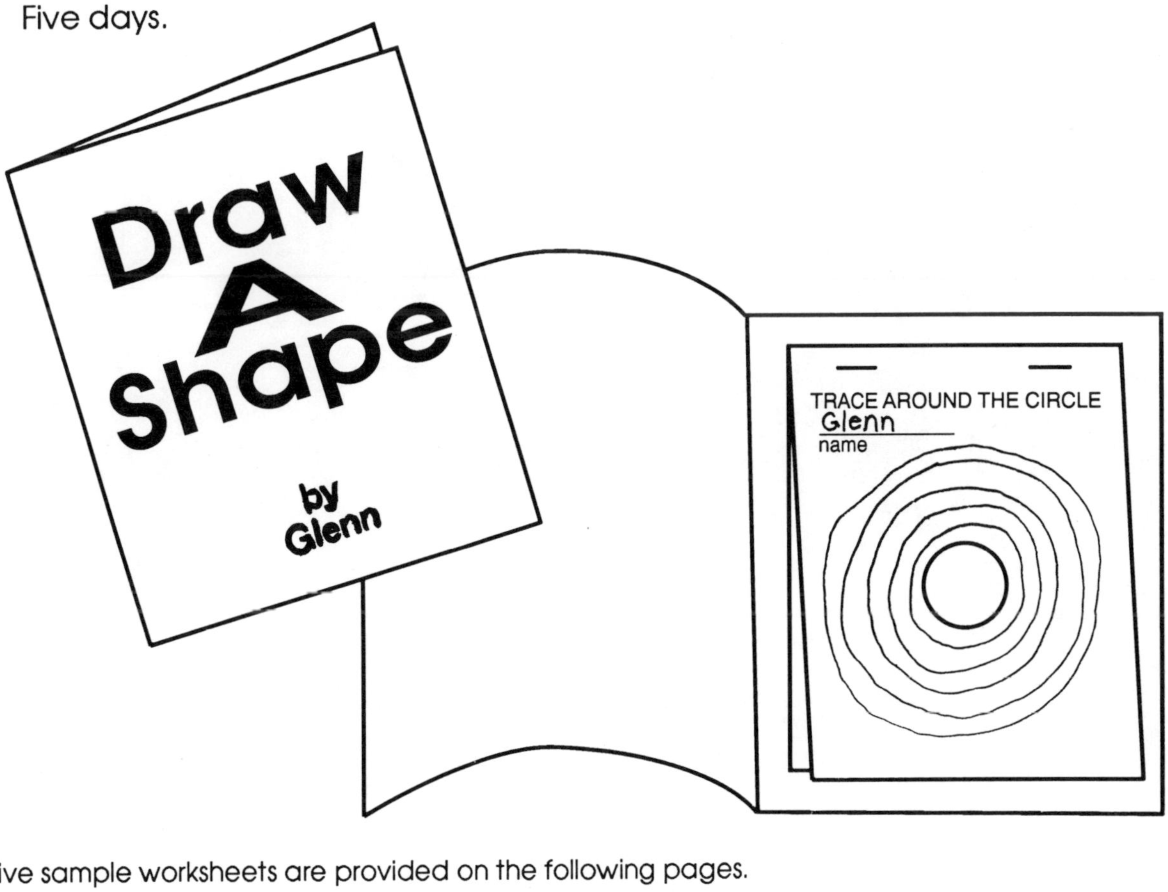

Five sample worksheets are provided on the following pages.

TRACE AROUND THE CIRCLE

Name____

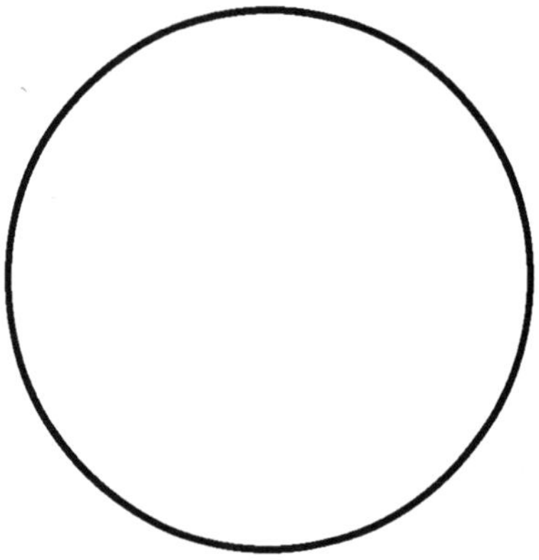

TRACE AROUND THE RECTANGLE

Name____

TRACE AROUND THE TRIANGLE

Name____

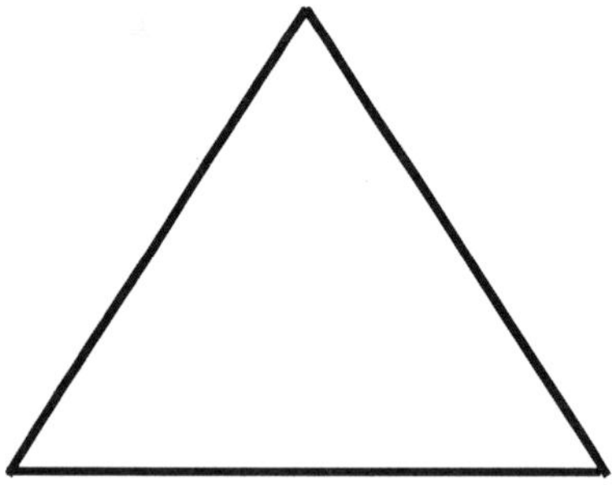

TRACE AROUND THE SQUARE

Name____

TRACE AROUND THE OVAL

Name____

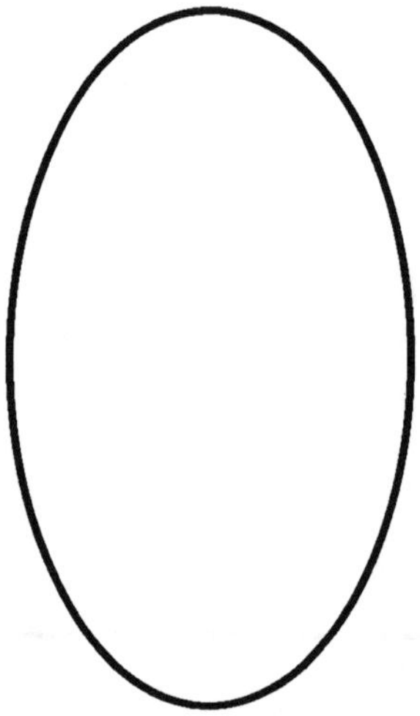

MY BOOK OF SQUARE THINGS

(My Book of Round Things, My Book of Triangular Things)

Description of Booklet

This booklet consists of a square cover that is constructed from two pieces of 9" x 9" colored construction paper. Center pages of typing paper are inserted before the booklet is stapled together along one edge.

Learning Skills Involved

Creativity and eye-hand coordination.

Method of Preparation

Booklets are prepared in advance by the teacher. Children are asked to draw square things on each page. For example, the subject on each page should differ and might include:

1. square people
2. square animals
3. square cars or trains or airplanes
4. square houses with square trees and square flowers
5. square toys

Preparation Time

As many days as suggested subjects.

ORDERED PAIRS

Description of Booklet

This booklet consists of a 12" x 18" cover of colored construction paper that is folded in half. Four or five worksheets containing grids are stapled inside the cover.

Learning Skills Involved

Eye-hand coordination and analytical thinking.

Method of preparation

The teacher constructs the booklets in advance. A sample grid is given on the following page.

Students place colored dots at each intersection, then using a ruler they connect the dots to make a geometric shape. When reading ordered pairs, the first number is placed from left to right (horizontally on the grid), while the second number tells how many units to move up.

Preparation Time

One grid per day. The following five ordered pairs may be used:
square – (3, 9,), (7, 9), (3, 5), (7, 5)
triangle – (5, 9), (2, 4), (8, 4)
parallelogram – (4, 10), (8, 10), (2, 7), (6, 7)
rectangle – (3, 10), (7, 10), (3, 3), (7, 3)
star – (5, 9), (4, 2), (8,4) then (5, 2), (8, 7), (2, 7)

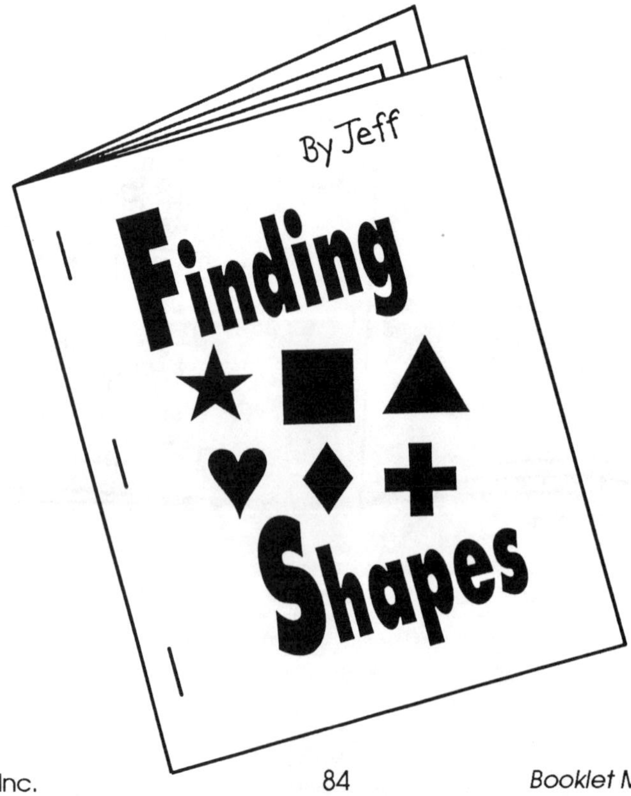

GRID FOR ORDERED PAIRS

Name____

Put a ______________ dot at each intersection: ________________________
(color)

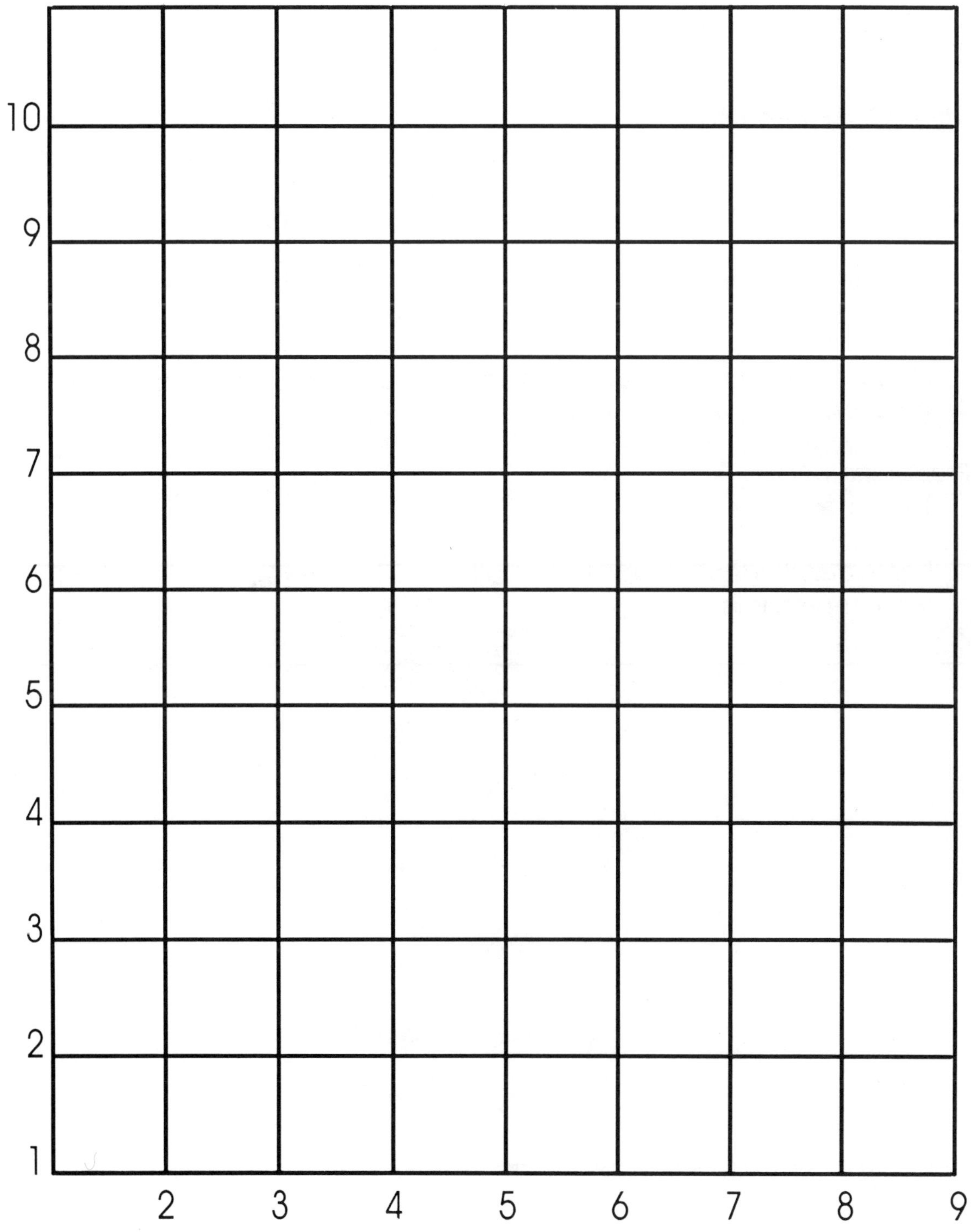

AUTUMN MONTHS

(Winter Months or Spring Months)

Description of Booklet

This booklet consists of a 12" x 18" cover of colored construction paper that is folded in half. Three inner worksheets are stapled inside.

Learning Skills Involved

Observation and eye-hand coordination.

Method of Preparation

Booklets are constructed in advance by the teacher. If an Autumn booklet is to be prepared; several October, November, and December calendars are placed on the work table for students to copy.

In Winter; January, February and March calendars are offered. In Spring; April, May, and June calendars are offered.

Students are asked to draw and color a picture at the top of each month's calendar.

Preparation Time

Three days.

Reminder: Students should be asked to observe on what day the first of each month falls.

Note: A calendar to be filled in is given on the following page.

MONTH ______________________________

Name ______________________________

Draw a picture

Sunday	Monday	Tuesday	Wednesday	Thursday	Friday	Saturday

NURSERY RHYME NUMBERS

Description of Booklet

This booklet consists of a 12" x 18" cover of colored construction paper that is folded in half. Several inner pages are inserted and stapled in place.

Learning Skills Involved

Listening comprehension and eye-hand coordination.

Method of Preparation

Booklets are constructed in advance by the teacher, one booklet per child. The teacher suggests a rhyme and then students and teacher repeat it together.

Children are given paper, pencil, and crayons and they draw the number of items suggested in the rhyme.

Preparation Time

Five or more days, depending on the number of nursery rhymes used. One rhyme per day is suggested. On the following page, several rhymes are listed.

Suggested rhymes for

NURSERY RHYME NUMBERS BOOKLET

Page 1: **Three Little Kittens**
The three little kittens
They lost their mittens and they began to cry.
Oh mother dear, see here, see here
Our mittens we have lost.
What, lost your mittens? You bad little kittens.
Then you shall have no pie.
Meow, meow, you shall have no pie.
(Draw the three little kittens.)

Page 2: **Two Little Blackbirds**
Two little blackbirds sitting on a hill
One named Jack, the other named Jill.
Fly away Jack! Fly away Jill!
Come back Jack! Come back Jill!
(Draw the two blackbirds.)

Page 3: **Three Men In A Tub**
Rub-a-dub-dub, three men in a tub.
And who do you think they be?
The butcher, the baker, the candle-stick maker,
Turn 'em out, knaves all three.
(Draw the three men in the tub.)

Page 4: **Baa, Baa Black Sheep**
Baa, baa black sheep have you any wool?
Yes, sir, yes sir, three bags full.
One for my master and one for my dame but
None for the little boy who lives down the lane.
(Draw the three bags full.)

Page 5: **Ten Little Indians**
One little, two little, three little Indians
Four little, five little, six little Indians
Seven little, eight little, nine little Indians,
Ten little Indian braves.
(Draw the 10 little Indian braves.)

Other suggestions:
Sing a Song of Sixpence (Draw the 24 blackbirds.)
The Old woman in the Shoe (Draw "so many children.")
Ole King Cole (Draw the fiddlers three.)
Wee Willie Winkie (Draw "past eight" on a clock.)

FILL IN THE TIME

Description of Booklet

This booklet consists of a 12" x 18" cover of colored construction paper that is folded in half. Several inner pages containing a clock face – minus hands – are stapled inside the cover after students have completed working on them.

Learning Skills Involved

Telling time.

Method of Preparation

Worksheets are prepared in advance by the teacher. Using the sample worksheet given on the following page, the subject is filled in for each. For example, the subject for the pages should differ and could include:

1. the time I get up
2. the time school begins
3. lunch time
4. the time school is over
5. dinner time
6. bedtime

Students draw the hands on the clock as indicated by the subject.

Preparation Time

From three to six days, depending on the clock worksheets used.

FILL IN THE TIME

Name____

Subject ____

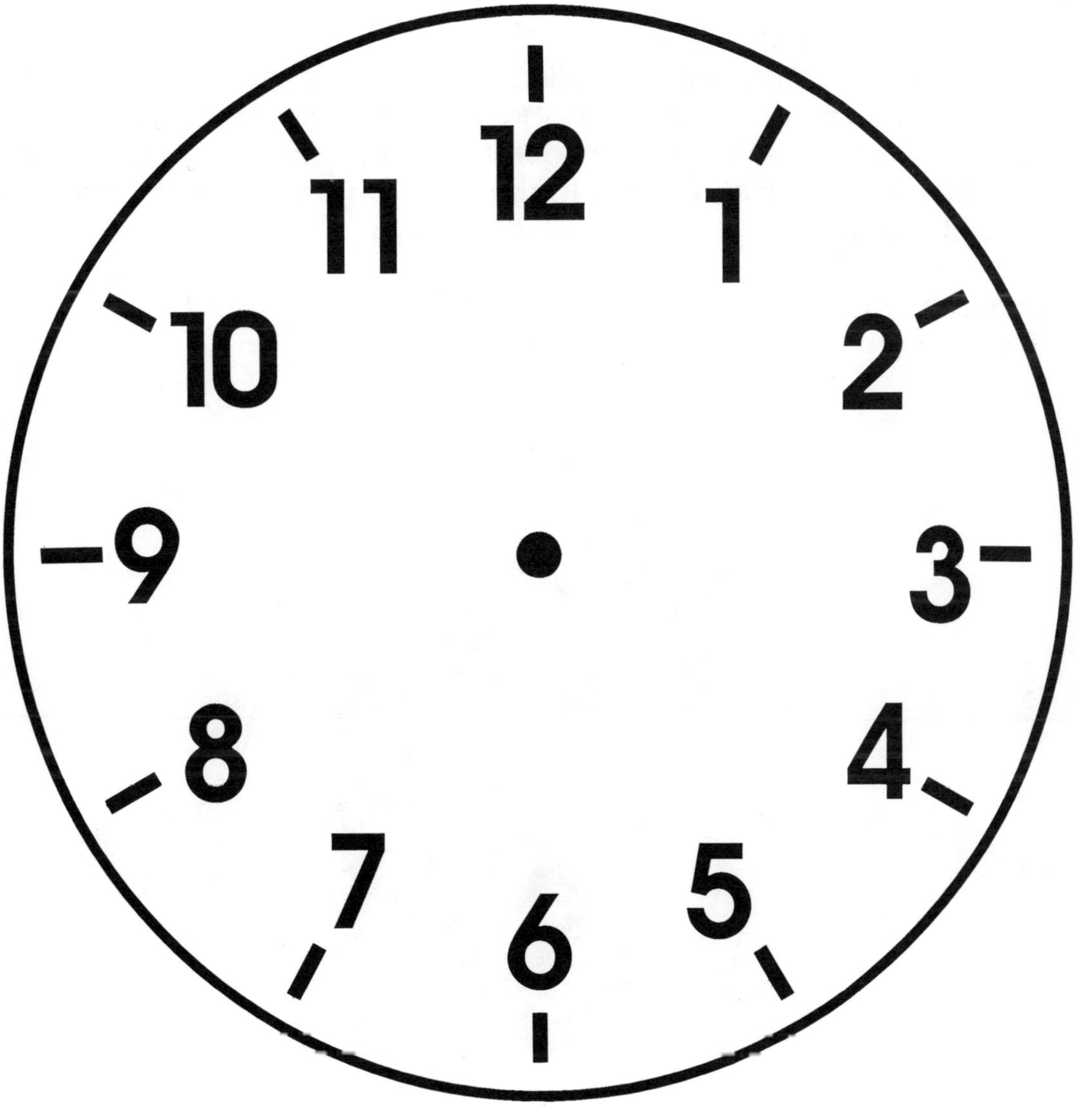

Time ____

COUNT TO 100

Description of Booklet

This booklet consists of a 12" x 18" cover of colored construction paper that is folded in half. Five lined inner pages are stapled inside the cover.

Learning skills Involved

Eye-hand coordination and counting.

Method of Preparation

Booklets are constructed in advance by the teacher. On the first page students write to 100 by ones. On the second page students write to 100 by twos. On page three, students write to – or as close to – 100 by threes. On page four, students write to 100 by fours. On page five, students write to 100 by fives.

Preparation Time

Five days.

FRACTIONS

Description of Booklet

This booklet consists of a 12" x 18" cover of colored construction paper that is folded in half. Three worksheets will be stapled inside the cover upon completion by students.

Learning Skills Involved

Visual discrimination.

Method of Preparation

Worksheets are prepared in advance by the teacher. Students are given crayons and asked to color 1/2, 1/3, and 1/4 of the geometric figures in the worksheets. Worksheets are given on the following three pages.

Preparation Time

Three days.

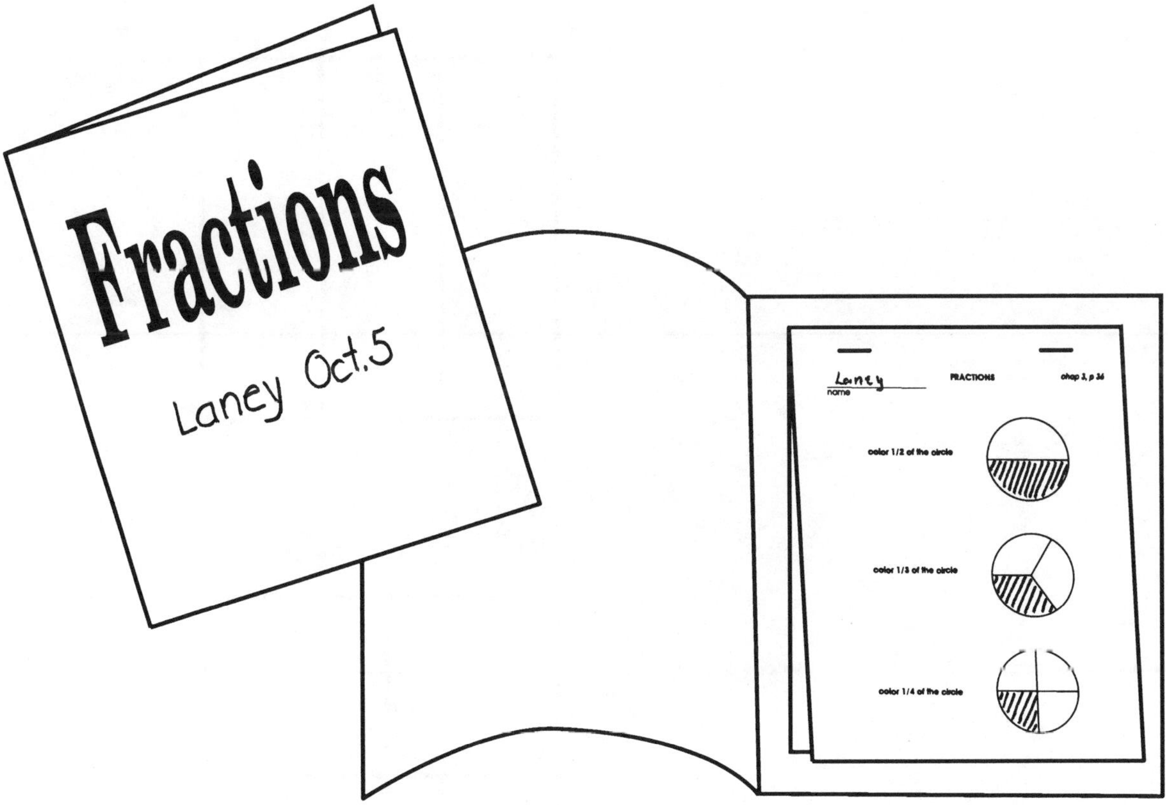

Note: Additonal worksheets using 2/3, 3/4, etc., may be prepared for older students.

FRACTIONS

Name____

Color 1/2 of the square

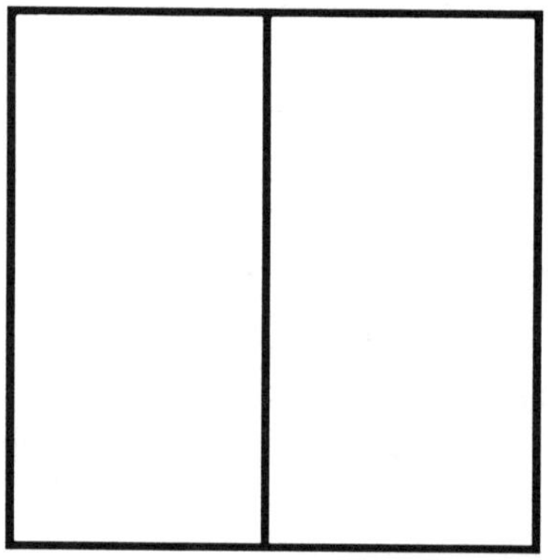

Color 1/3 of the square

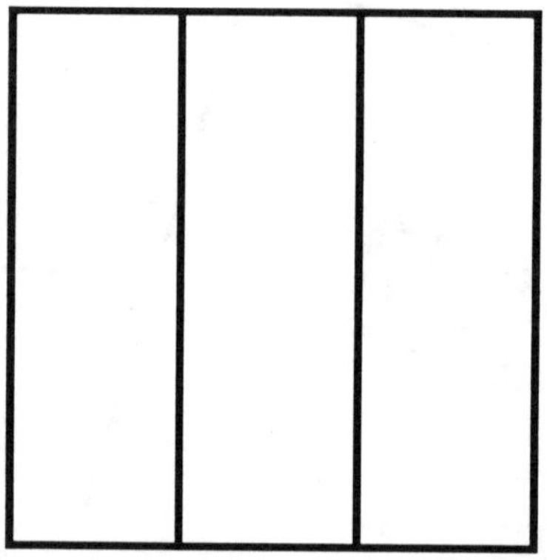

Color 1/4 of the square

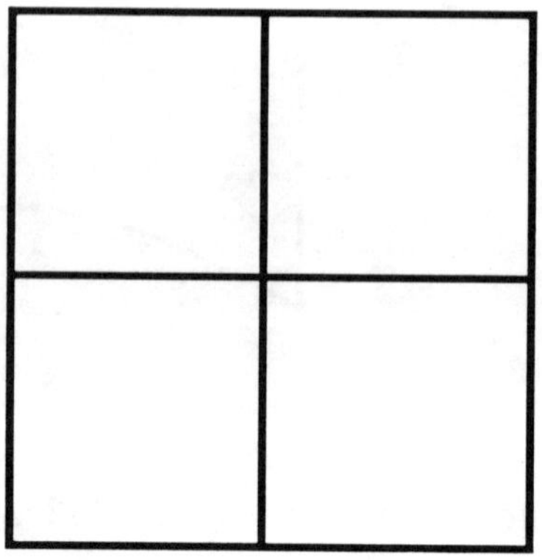

FRACTIONS

Name_______________________________________

Color 1/2 of the circle

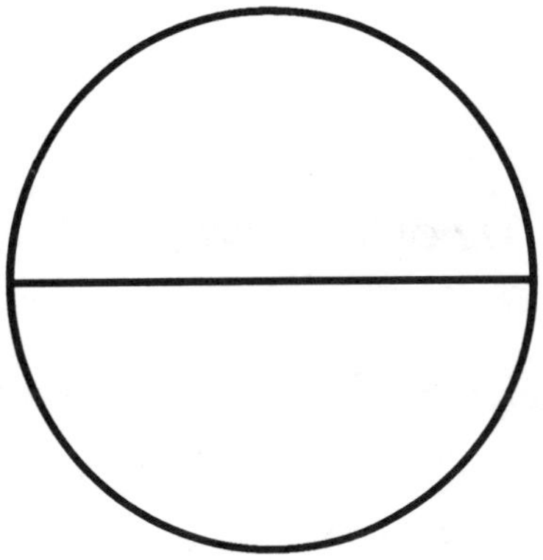

Color 1/3 of the circle

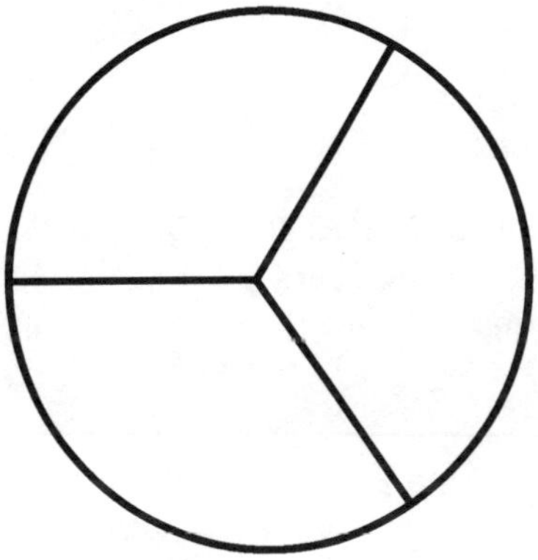

Color 1/4 of the circle

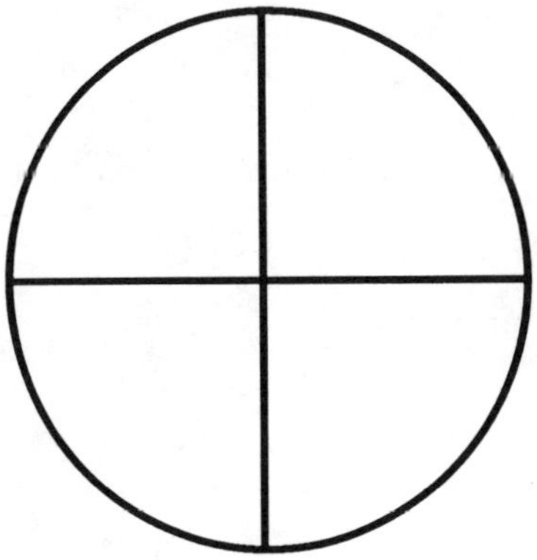

FRACTIONS

Name____

Color 1/2 of the triangle

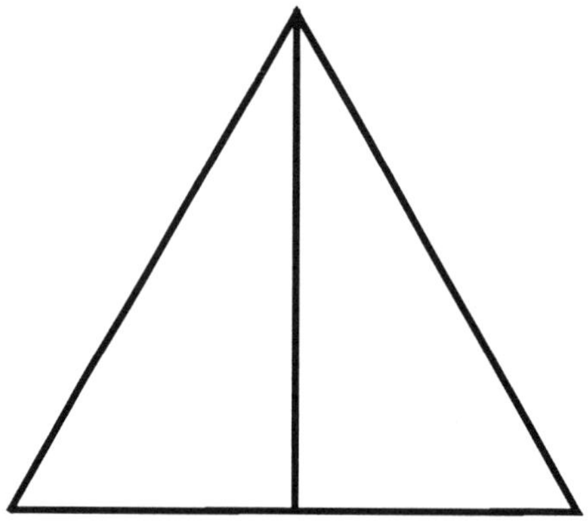

Color 1/3 of the triangle

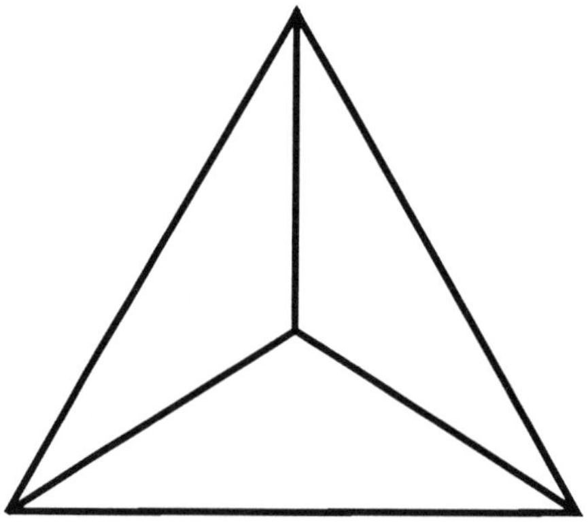

Color 1/4 of the triangle

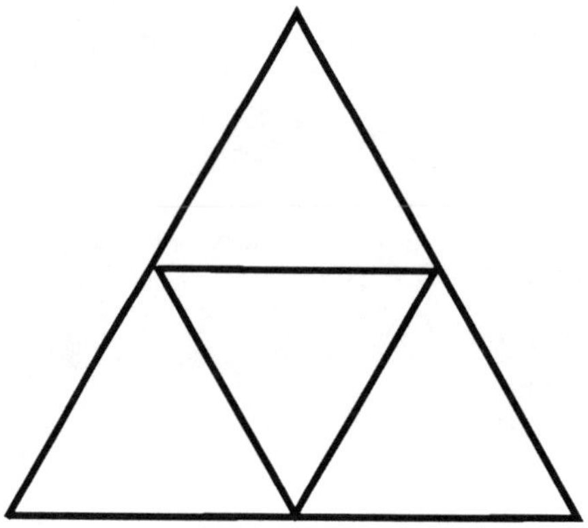

MEASURING

Description of Booklet

This booklet consists of a 12" x 18" cover of colored construction paper that is folded in half. Several worksheets are stapled inside the cover after students have completed them.

Learning Skills Involved

Visual discrimination.

Method of Preparation

Worksheets are prepared in advance by the teacher. They consist of patterns or shapes which students are asked to measure:

1. the tree
2 the boat
3. the see-saw

Pencils and rulers are provided.

Preparation Time

Three days or more, depending on the number of worksheets.

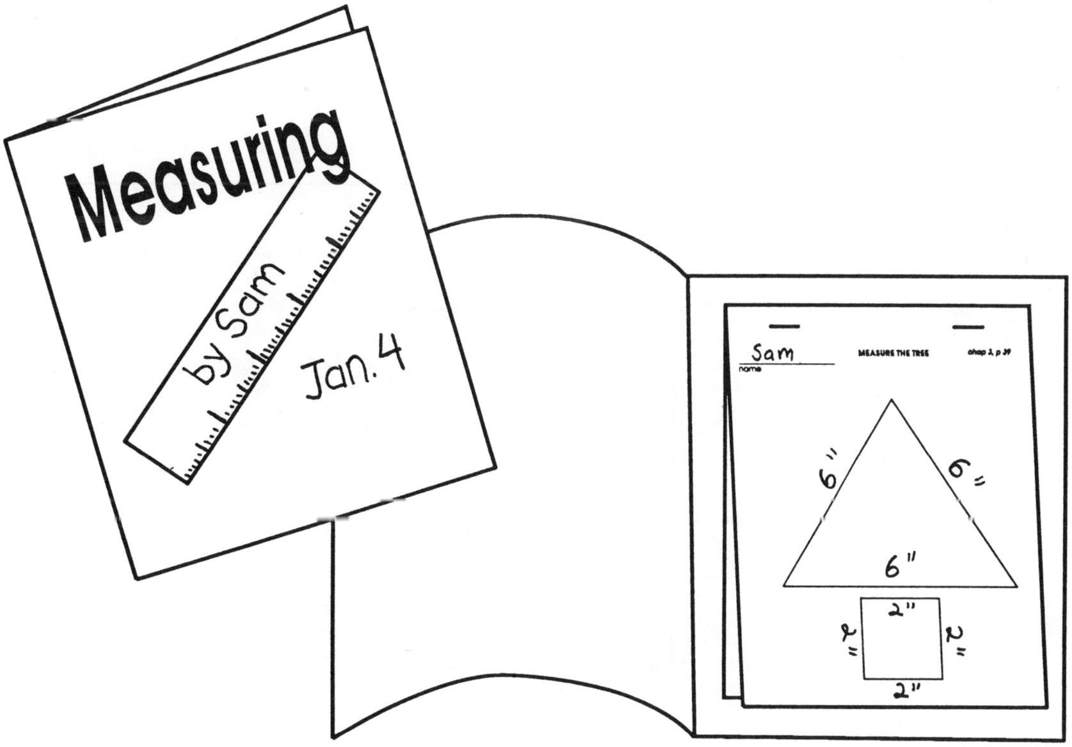

Three examples of worksheets are given on the following pages.

MEASURE THE TREE

Name____

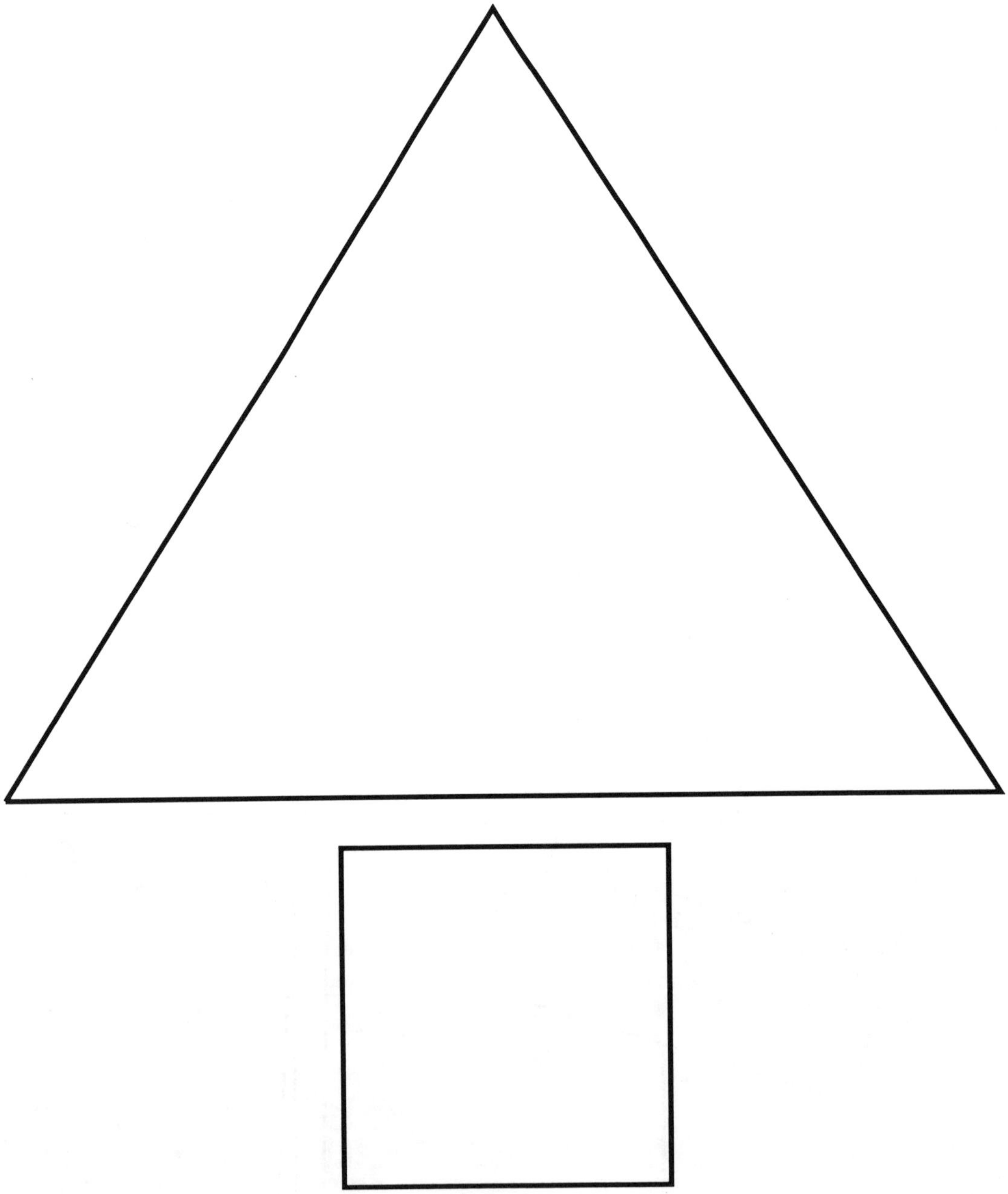

MEASURE THE BOAT

Name____

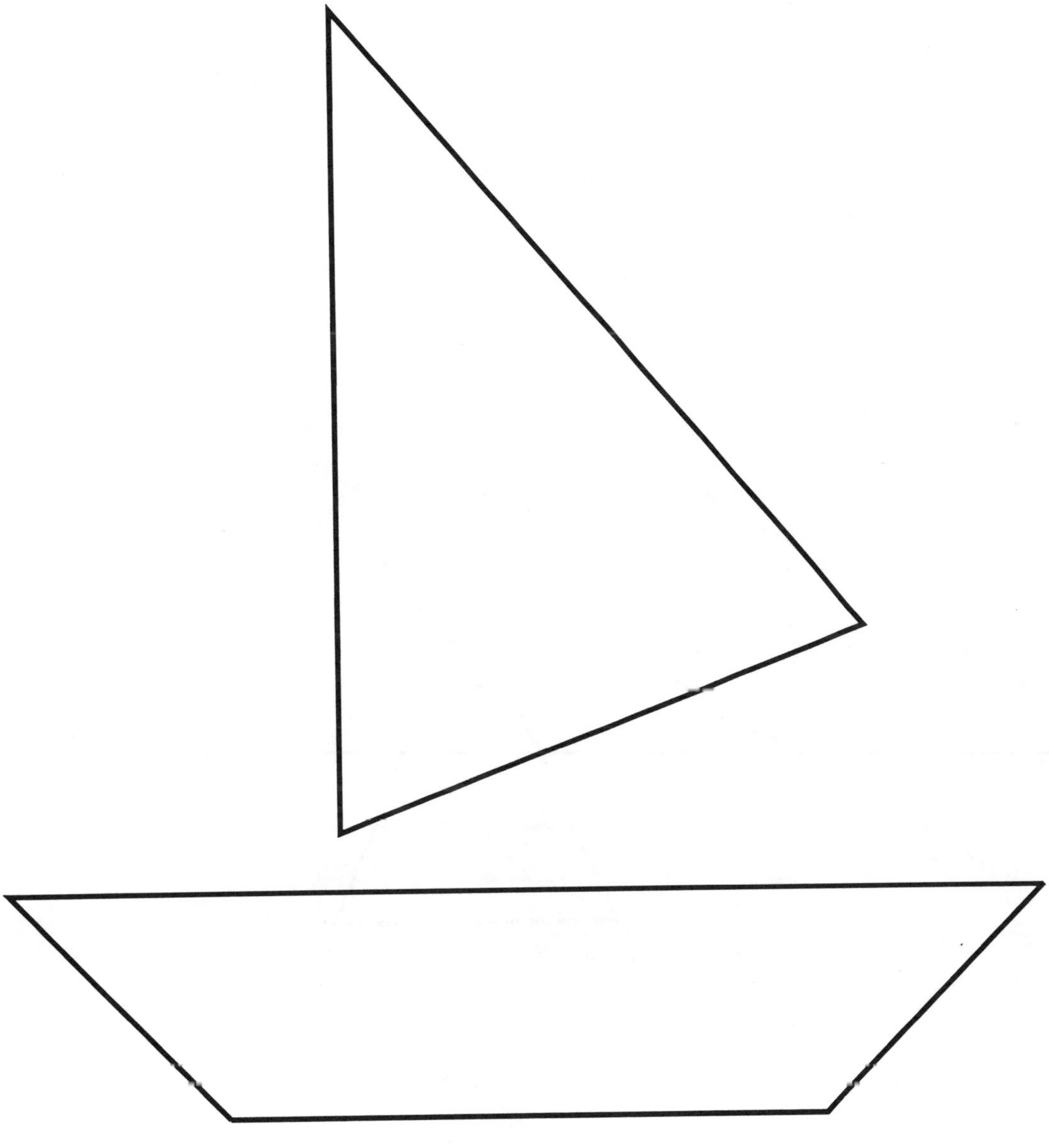

MEASURE THE SEE-SAW

Name____

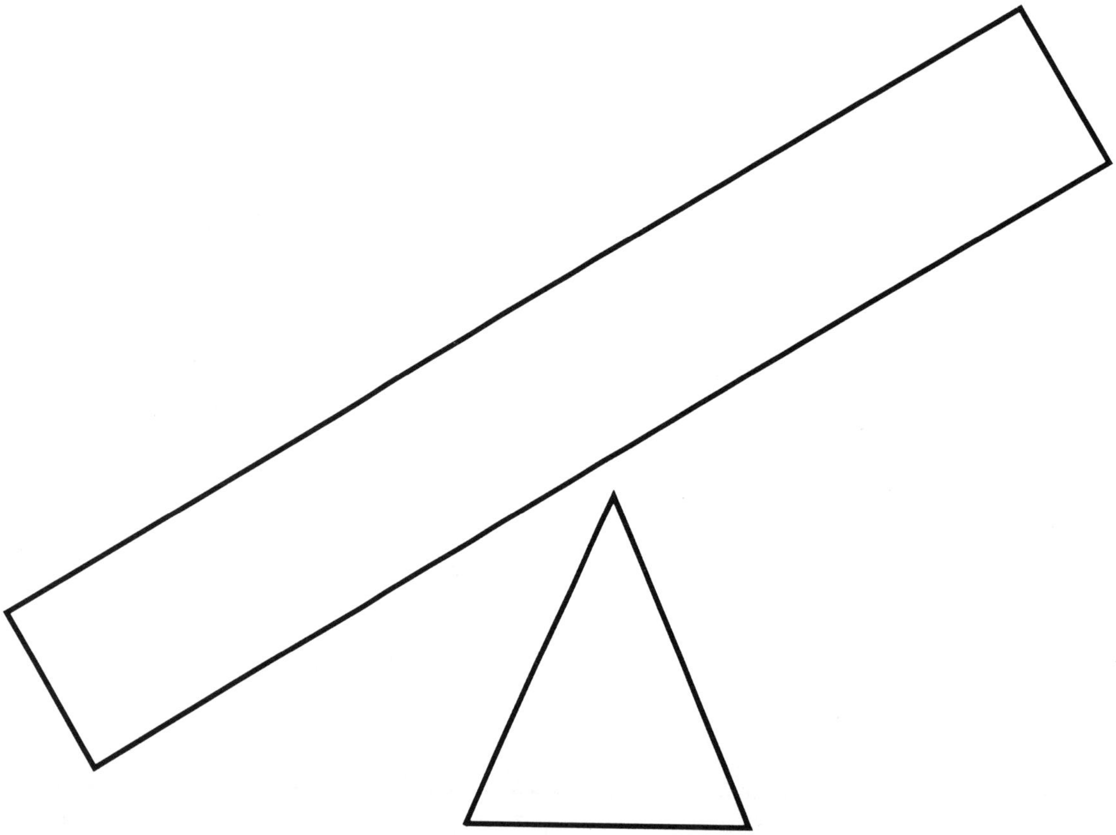

Chapter 4

SCIENCE

The activities in this chapter are planned to enhance the following topics:

- the five senses
- plants
- animals
- seasons and weather
- geography
- outer space
- scientific observation

I LIKE TO . . .

(Explaining the five senses)

Description of Booklet

This booklet consists of a 12" x 18" cover of colored construction paper that is folded in half. Five inner pages are stapled inside the cover. At the top of each inner page, the name of one of the senses is written or a simple illustration is drawn.

Learning Skills Involved

Imagination, observation and eye-hand coordination.

Method of Preparation

Either students or the teacher place an illustration or a title at the top of each page. Students then draw or write about things they like to see, hear, feel, taste or touch on the appropriate page.

Preparation Time

Five days.

Examples of illustrations are given on the following page.

Examples of illustrations for

I LIKE TO . . .

(These are to be cut out and pasted at the top of the pages.)

I LIKE TO 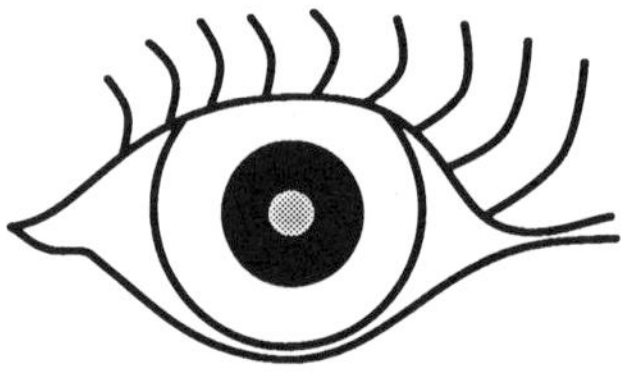

I LIKE TO 

I LIKE TO 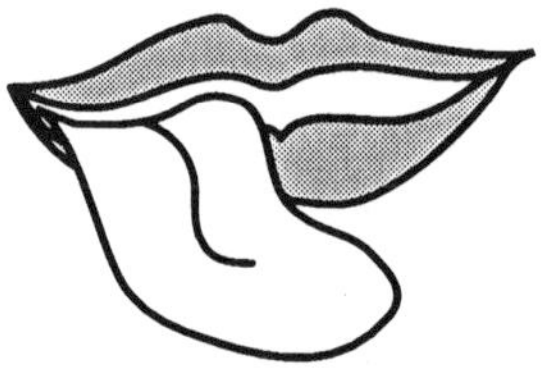

I LIKE TO 

I LIKE TO 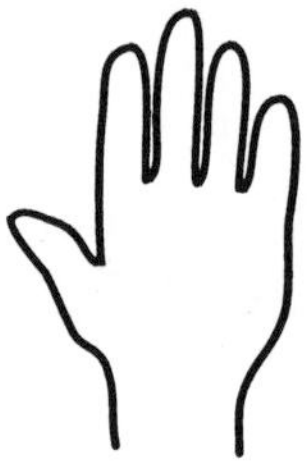

Sight

MIXING COLORS

Description of Booklet

This booklet consists of a 12" x 18" cover of colored construction paper that is folded in half. Six inner pages, each with a large circle duplicated in the center, are stapled to the back cover as they are completed by students.

Learning Skills Involved

Eye-hand coordination and observation.

Method of Preparation

Beginning with the primary colors, students paint a circle red on the first day, yellow on the second day and blue on the third day.

Secondary colors will be mixed by students on the next three days:

1. bowls of red and yellow paint are placed on the work table along with empty bowls for mixing. Each student combines a little red with yellow and paints a circle orange.
2. the same system is repeated the next day using yellow and blue to obtain a green circle.
3. the same system is repeated the last day using blue and red to obtain a purple circle.

Preparation Time

Three days.

Suggestion: Each page is labeled a different color. Also, students should clean up after painting.

Touch

MY FEELIE BOOK

Description of Booklet

This booklet consists of a 9" x 12" cover of colored construction paper that is folded in half. Several inner pages of typing paper or tablet paper are stapled inside the folder.

Learning Skills Involved

Sensory/touch and observation.

Method of Preparation

A variety of multi-sensory material is placed on the work table along with scissors, glue and pencils. Material might include: burlap, cellophane, silk, velvet, sandpaper, aluminum foil, sponges, rug samples, cotton balls, sticky tape, dotted swiss, etc.

Students glue one piece of material on each page. A list of the various materials may be placed on the work table and children can label each page.

Preparation Time

One day.

Taste

FOUR FOOD GROUPS

Description of Booklet

This booklet consists of a 12" x 18" cover of colored construction paper that is folded in half. Four inner pages of typing paper are stapled inside the cover.

Lesarning Skills Involved

Observation and eye-hand coordination.

Method of Preparation

Booklets are prepared in advance by the teacher. Pages are labeled. Children are given old magazines and workbooks, scissors and glue. They search for pictures of foods, cut them out, and paste them on the proper page of their booklets.

Preparation Time

Two days.

An example of the four pages is given on the following page.

Example pages for

FOOD GROUPS

page 1

fruits & vegetables

page 2

meat, fish, & poultry

page 3

cereal & bread

page 4

milk, cheese, & butter

Taste

MY BOOK OF VEGETABLES

Description of booklet

This booklet consists of a 9" x 12" cover of colored construction paper that is folded in half. Several pages of lined tablet paper are stapled to the inner back cover.

Learning Skills Involved

Eye-hand coordination and observation.

Method of Preparation

Booklets are prepared in advance by the teacher. A large chart showing various vegetable categories is placed on the work table.

Each page of the booklet is labeled and it focuses on one category of vegetables such as:

leaves, flowers, roots, tubers, stems or seeds

Children are given pencils, crayons and/or felt-tipped markers. They may copy all or part of each list as shown on the vegetable chart or they may draw various vegetables on their papers. (Pictures of vegetables may also be displayed.)

Preparation Time

Two or three days.

A vegetable chart is given on the following page.

Taste

VEGETABLES

LEAVES	FLOWERS	ROOTS	TUBERS	STEMS	SEEDS
chicory	broccoli	beets	potatoes	broccoli	peas
kale	cauliflower	onions	sweet potatoes	asparagus	string beans
endive	artichoke	carrots	peanuts	rhubarb	corn
watercress		parsnips	artichokes	celery	eggplant
dandelion		turnips		kohlrabi	lima beans
swiss chard		radishes			wax beans
beet greens		rutabagas			okra
lettuce					
cabbage					
spinach					
celery					
parsley					
salsify					
chard					
					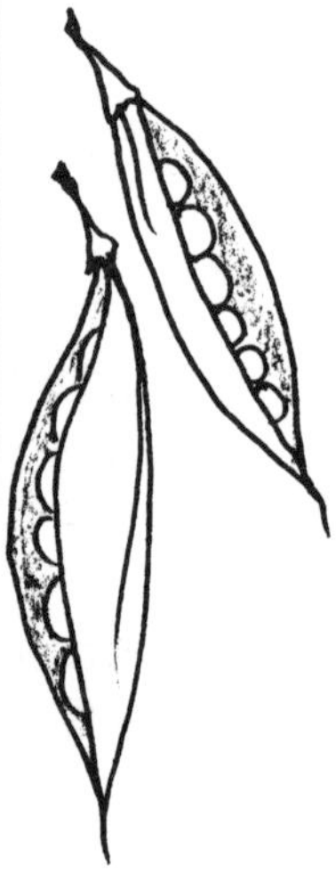

Taste

MY FAVORITE MEALS

Description of Booklet

This booklet consists of a 12" x 18" cover of colored construction paper that is folded in half. Three inner pages are stapled inside the cover.

Learning Skills Involved

Eye-hand coordination and personal judgement.

Method of Preparation

Pages are labeled: Breakfast, Lunch and Dinner. Booklets are prepared in advance by the teacher.

During discussion period, children are asked to name some of their favorite foods in each of the four categories. Illustrated charts of food groups should be displayed on or near the work table. Children are furnished with pencils and crayons and asked to draw their favorite foods for each meal and label each item. Teacher assistance for labeling may be needed by younger students.

Preparation Time

One or two days.

MY BOOK OF LEAVES

Description of Booklet

This booklet consists of a 12" x 18" cover of colored construction paper that is folded in half. Inner pages of typing paper will be stapled inside the cover upon completion.

Learning Skills Involved

Eye-hand coordination and visual identification.

Method of Preparation

Each day the teacher brings leaves from one particular species of tree to the classroom. The child selects a leaf, rib side up, and positions a sheet of typing paper over it. Then using the side of a "peeled" crayon, he/she rubs it. More than one leaf may be "rubbed" on a paper.

The name of the tree is placed prominently on a card in the center of the work table. Children copy the name on their papers.

Preparation Time

Five days.

TREES

Description of Booklet

This booklet consists of a 12" x 18" cover of colored construction paper that is folded in half. Four inner pages of typing paper are stapled inside.

Learning Skills Involved

Eye-hand coordination and observation.

Method of Preparation

Booklets are prepared in advance by the teacher. Students work on one page a day. The following suggestions for each day's work may be used:

1. Name and draw one or more trees in your neighborhood.
2. Name the parts of a tree on a duplicated pattern. (An example is given on the following page.)
3. Draw or write how trees are helpful to people. For example: they give us shade, wood for fires, clean air, hold the soil, a place for birds and squirrels to build homes, etc.
4. Draw products we get from trees.

Class discussions should precede these activities.

Preparation Time

Four days.

Suggestions for this booklet appear on the following pages.

Sample pages for

TREES

page 1

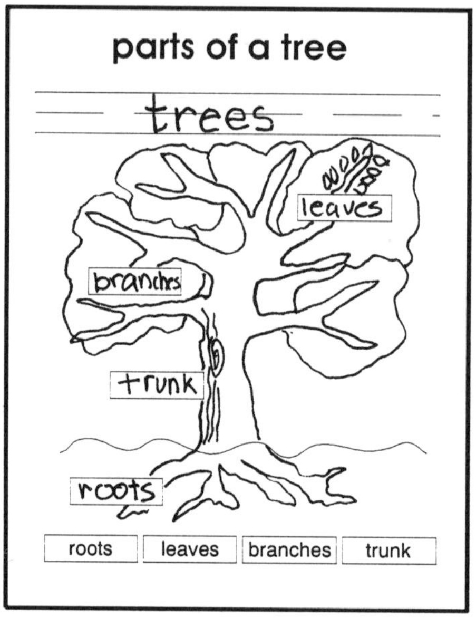

page 2

page 3

page 4

TREES

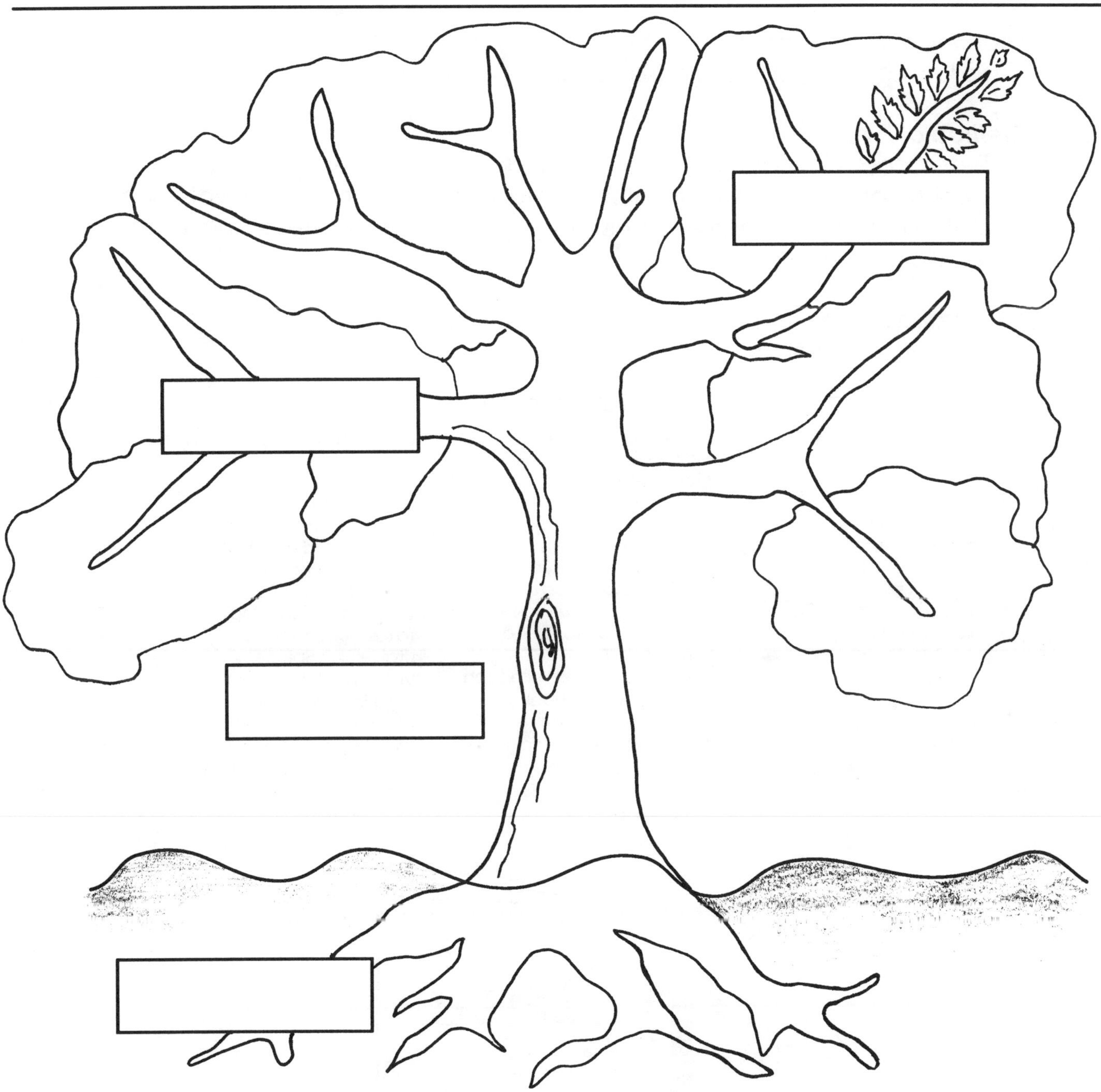

roots	leaves	branches	trunk

STATE TREES

Description of Booklet (Group Booklet)

This booklet consists of a 12" x 18" cover of two pieces of oak tag or poster board. When all work has been completed, inner pages are stapled inside or brads may be used to hold the booklet together.

Learning Skills Involved

Following directions and observation.

Method of Preparation

Reference books are placed on the work table along with manila paper, pencils, crayons and/or felt-tipped markers. Each student selects one state and draws the tree on his/her paper. Names of trees and states should be included.

Preparation Time

Two or three days.

State Trees

Alabama – Southern Pine
Alaska – Sitka Spruce
Arizona – Paloverde
Arkansas – Pine
California – Redwood
Colorado – Colorado Blue Spruce
Connecticut – White Oak
Delaware – American Holly
Florida – Sabal Palmetto Palm
Georgia – Live Oak
Hawaii – Candlenut
Idaho – White Pine
Illinois – White Oak
Indiana – Tulip Popladr
Iowa – Oak
Kansas – Cottonwood
Kentucky – Kentucky Coffee Tree
Louisiana – Cypress
Maine – Easter White Pine
Maryland – White Oak
Massachusetts – American Elm
Michigan – White Pine
Minnesota – Norway Pine
Mississippi – Magnolia
Missouri – Dogwood
Montana – Ponderosa Pine
Nebraska – Cottonwood
Nevada – Piñon
New Hampshire – White Birch
New Jersey – Red Oak
New Mexico – Piñon
New York – Sugar Maple
North Carolina – Pine
North Dakota – American Elm
Ohio – Buckeye
Oklahoma – Redbud
Oregon – Douglas Fir
Pennsylvania – Hemlock
Rhode Island – Red Maple
South Carolina – Palmetto
South Dakota – Black Hills Spruce
Tennessee – Tulip Poplar
Texas – Pecan
Utah – Blue Spruce
Vermont – Sugar Maple
Virginia– Dogwood
Washington – Western Hemlock
West Virginia – Sugar Maple
Wisconsin – Sugar Maple
Wyoming – Cottonwood

ANIMALS

Description of Booklet

This booklet consists of a 12" x 18" cover of colored construction paper that is folded in half. Three inner pages are stapled inside the back cover along the fold.

Learning Skills Involved

Eye-hand coordination and memory.

Method of Preparation

Each of the pages is titled either: Pets, Farm Animals or Wild Animals. Students draw several animals in each category using the front and back of each page.

Preparation Time

Three days. Class discussion(s) prior to independent work is suggested.

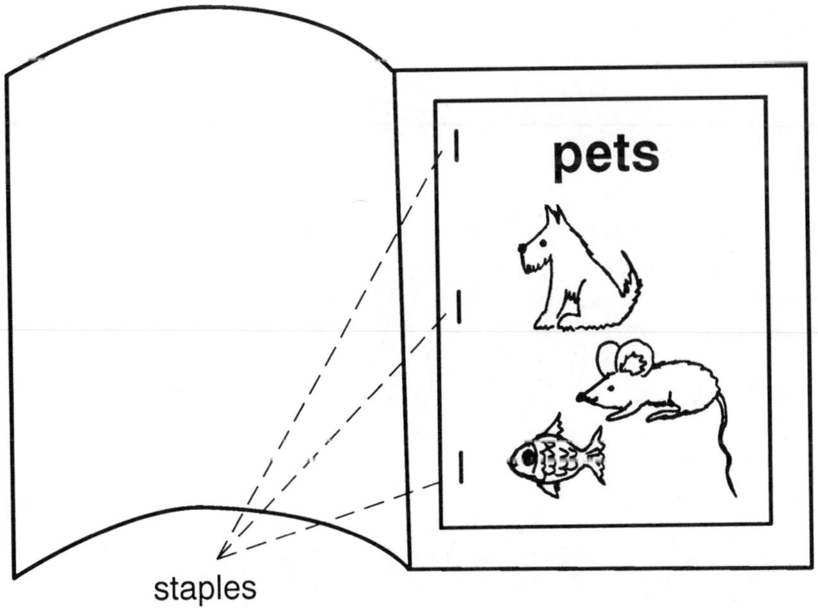

ANIMAL GROUPS

Description of Booklet

This booklet consists of a 12" x 18" cover of colored construction paper that is folded in half. Inner pages are stapled inside the cover when completed by students.

Learning Skills Involved

Listening comprehension, comparing, classifying and summarizing.

Method of Preparation

Several animal groups are selected by the teacher for study. Each day the class discusses one group. Students then prepare a page of their booklets relating to that day's discussion. The following groups may be considered: mammals, birds, fish, insects, amphibians, reptiles, etc.

Students may be shown pictures of one group each day to aid them in drawing, coloring and creating their own animals. Pencils, crayons, and/or colored markers should be provided.

Preparation Time

One day for each group of animals. Primary grade children may be asked to write one or two facts about each group on their papers such as: where they live, how they move, what noises they make, what kind of body covering they have, etc.

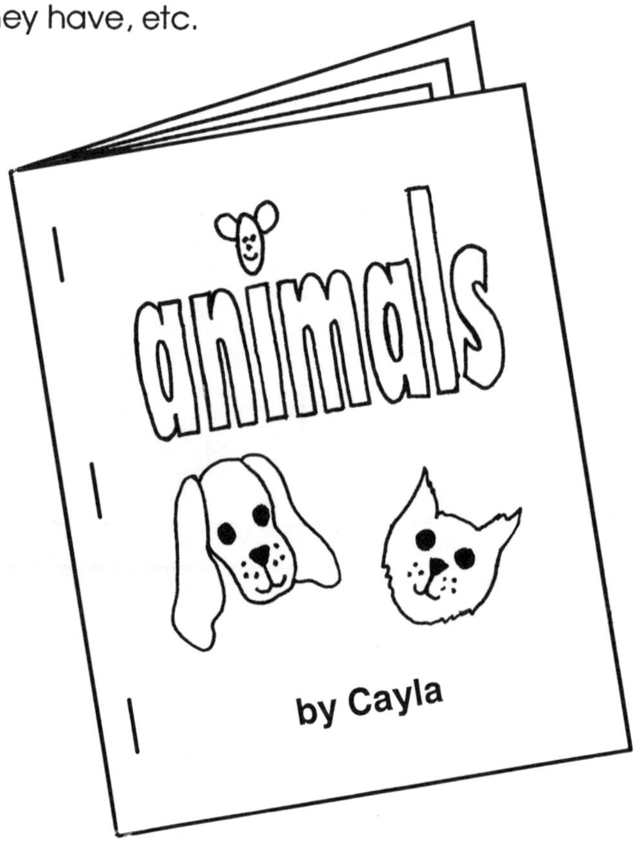

HOW ANIMALS MOVE

Description of Booklet

This booklet consists of a 12" x 18" cover of colored construction paper that is folded in half. Four or five titled inner pages are stapled inside the cover. Individual titles may include: run, hop, crawl, fly, swim, etc.

Learning Skills Involved

Eye-hand coordination and creativity.

Method of Preparation

Under the title the student draws one or more animals that fit the category.

Preparation Time

Two to five days.

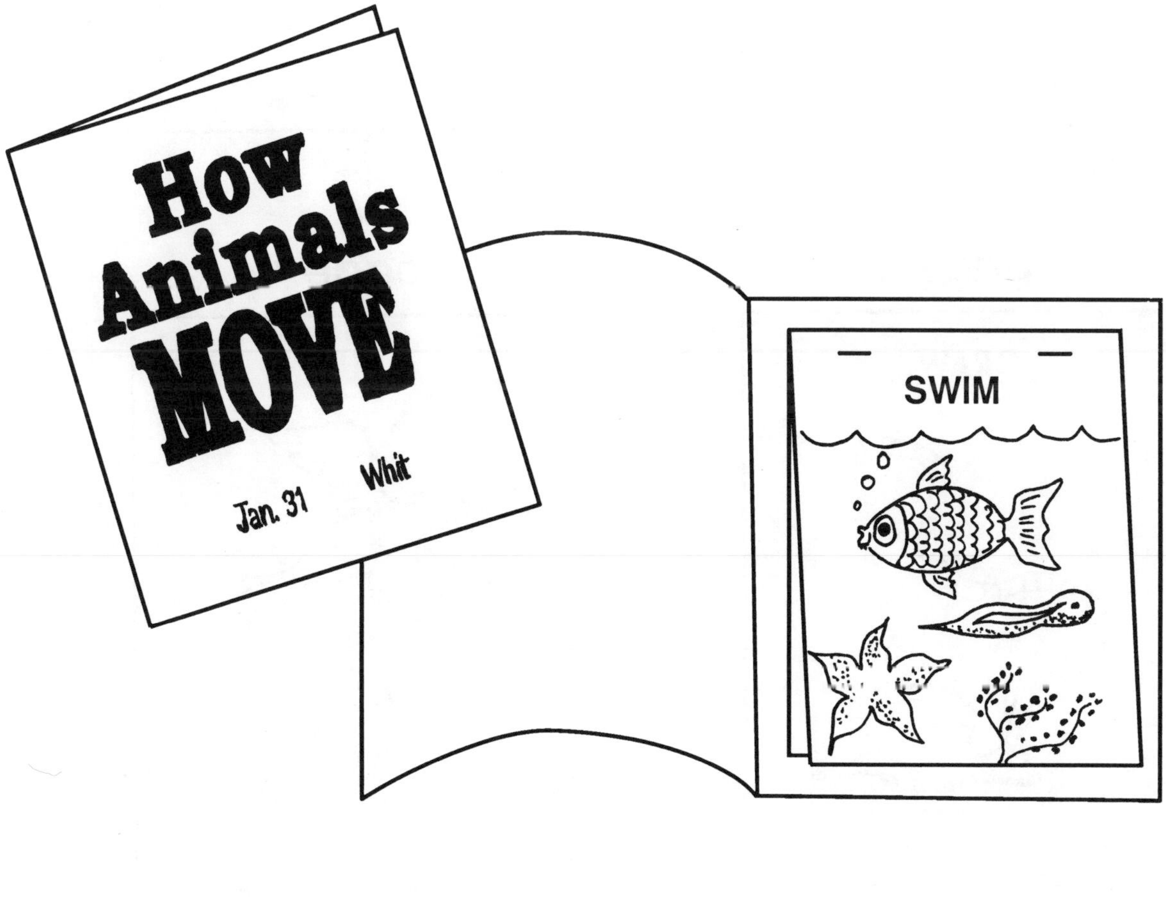

Ideas for this booklet are given on the following page.

HOW ANIMALS MOVE

Name____

page 1

RUN

page 2

HOP

page 3

CRAWL

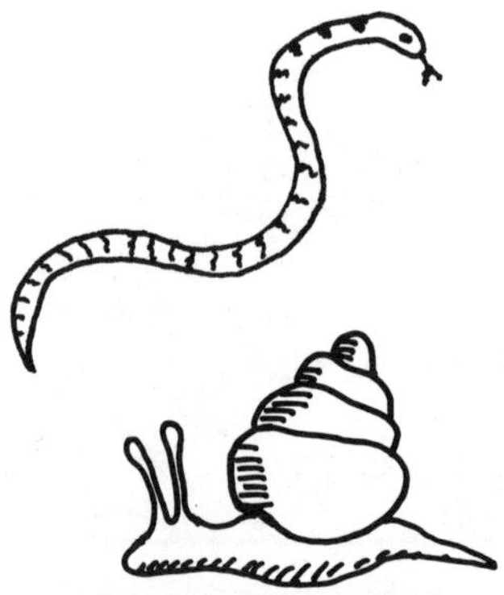

page 4

FLY

Class discussion(s) prior to independent work is suggested.

SNAKES

(Birds, Fish, or Insects)

Description of Booklet

This booklet consists of a 12" x 18" cover of colored construction paper that is folded in half. Inner pages are stapled to the back cover.

Learning Skills Involved

Observation and investigation.

Method of Preparation

Students are given a choice of investigating one of the above animal categories. Reference books should be placed on the work table along with pencils and crayons and/or felt-tipped markers. Students are encouraged to use the school library.

Examples of what may be required:

1. Draw a different snake each day, color and label it.
2. State if it is poisonous or non-poisonous.
3. Tell where it lives.
4. Describe how it might be helpful to environment.

Preparation Time

Four days.

BUTTERFLIES

Description of Booklet

This booklet consists of a cover of colored construction paper that is folded in half. Several inner pages are stapled inside the cover.

Learning Skills Involved

Eye-hand coordination and visual discrimination.

Method of Preparation

Booklets are prepared in advance by the teacher. Cardboard or posterboard patterns of butterflies are placed on the work table for students to trace. Optional: colorful pictures of butterflies may be placed on the work table to inspire students.

Pencils and colored marking pens are made available along with patterns. Students are encouraged to color butterflies imaginatively.

Preparation Time

One or two days.

Two butterfly patters are given on the following page.

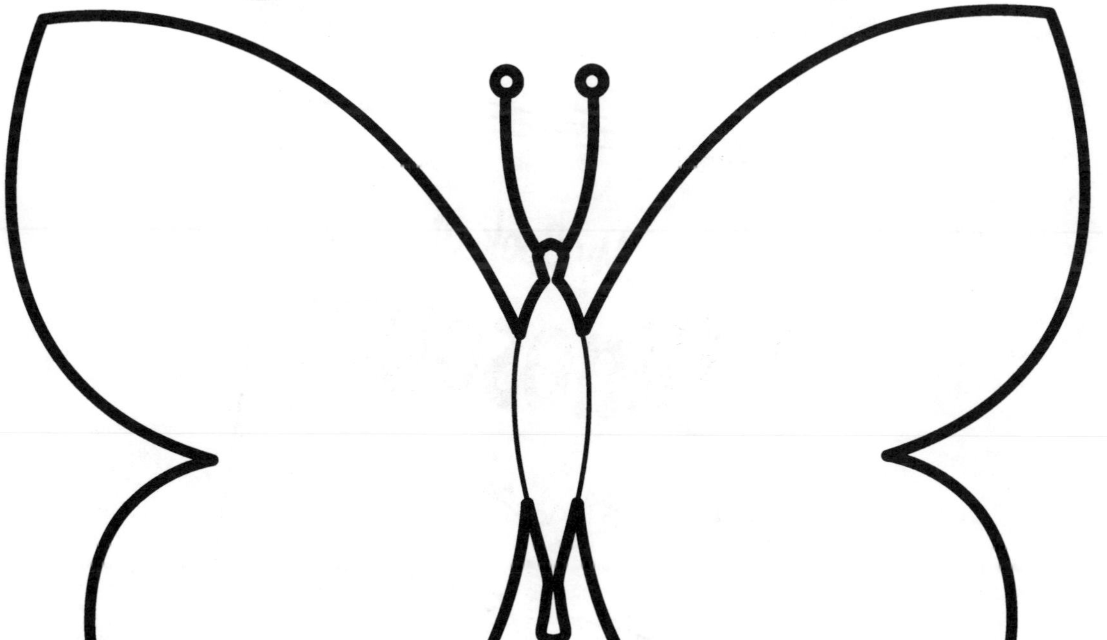

MY BOOK OF DINOSAURS

Description of Booklet

This booklet consists of a 12" x 18" cover of colored construction paper that is folded in half. Inner pages of typing paper are stapled inside the folder when completed by students.

Learning Skills Involved

Eye-hand coordination and imagination.

Method of Preparation

Dinosaur drawings for children to trace are placed on the worktable along with crayons/felt-tipped markers.

Class discussions of dinosaurs, fauna and type of terrain are suggested. Papers should reflect this information.

Preparation Time

Three to five days.

BRONTOSAURS

STEGOSAURUS

ANKYLOSAURUS

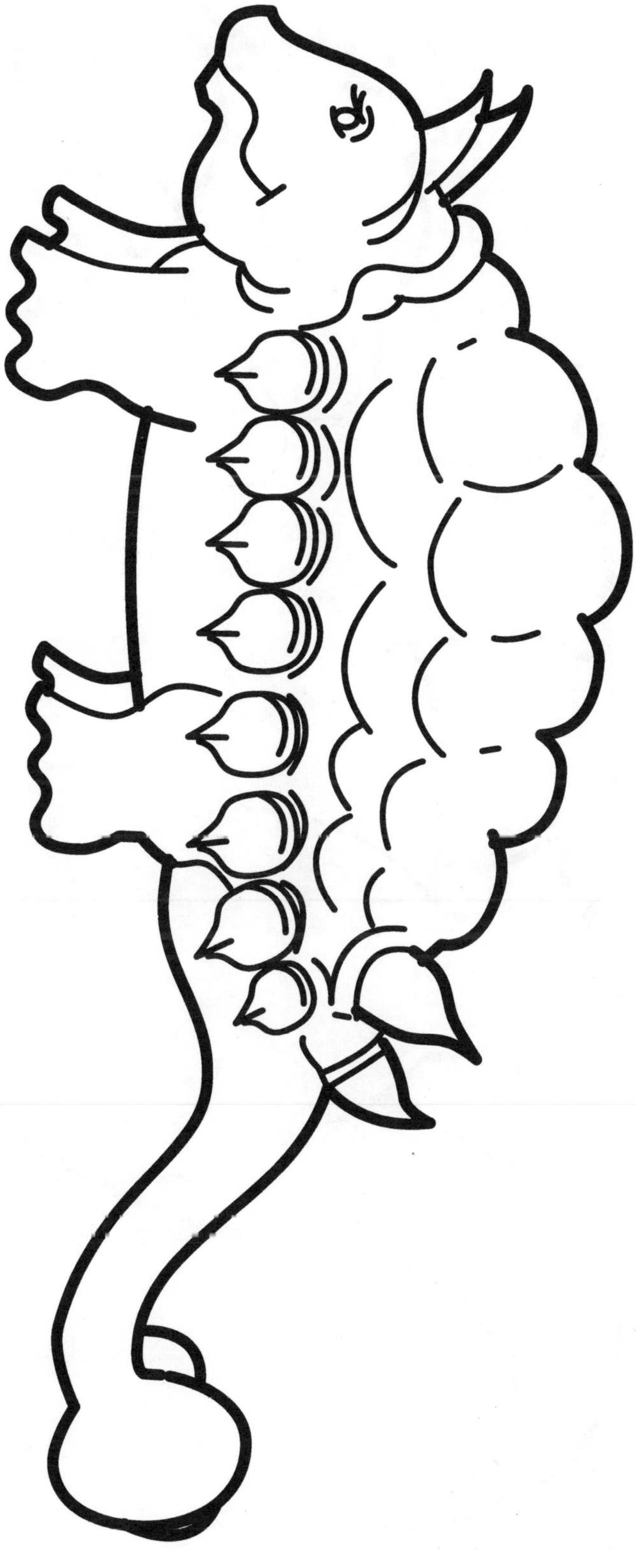

TRICERATOPS

TYRANNOSAURUS REX

PTERANODON

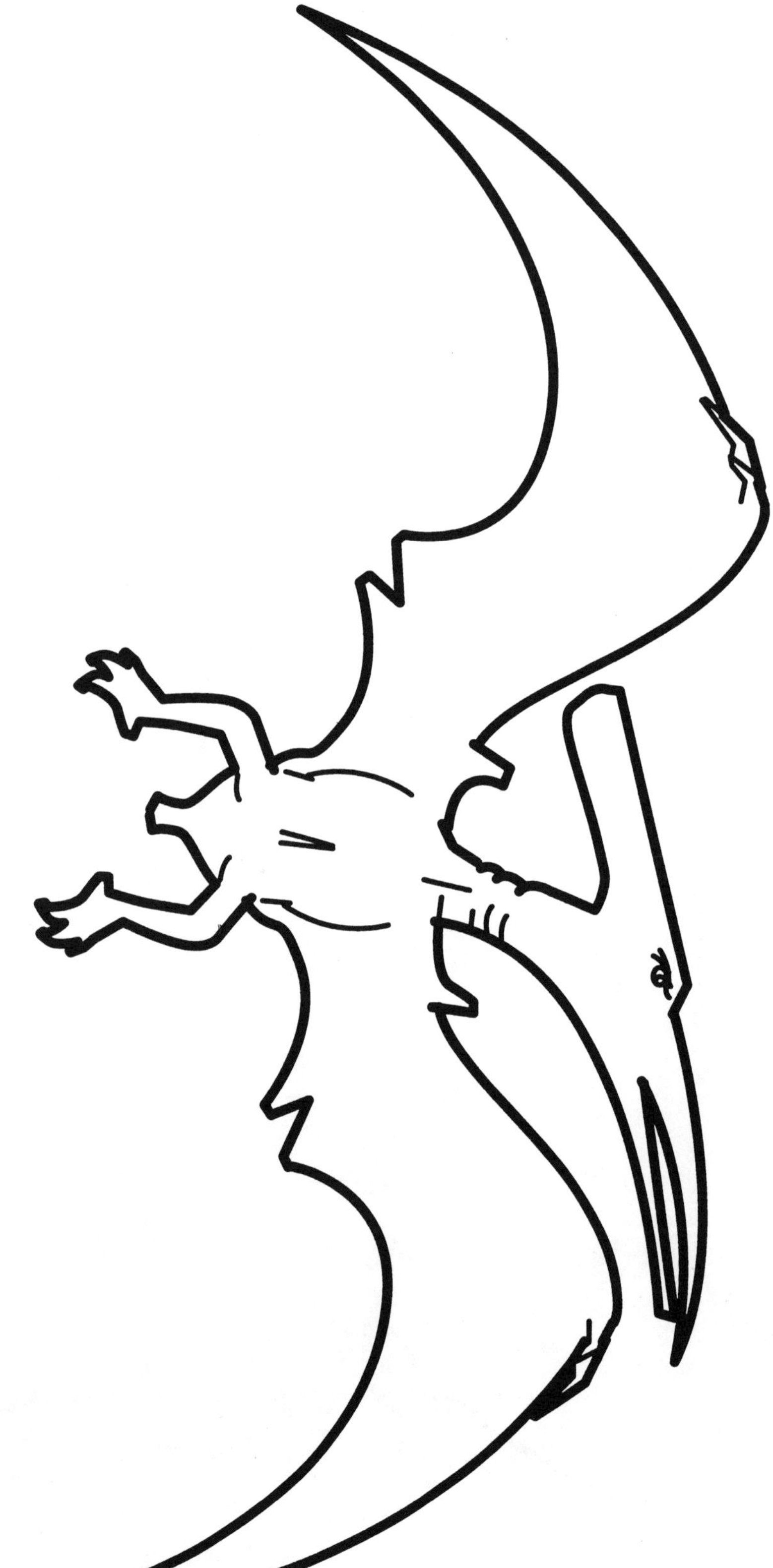

WEATHER

Description of Booklet

This booklet consists of a 12" x 18" cover of colored construction paper that is folded in half. Typing paper is used for inner pages that are stapled inside the cover as completed by students.

Learning Skills Involved

Eye-hand coordination and imagination.

Method of Preparation

Each day during discussion period one aspect of weather is considered such as rainy, sunny, snowy, cloudy, windy. Children are given paper, pencils and crayons and asked to draw a picture that illustrates the given topic. Papers are labeled and collected by the teacher. At the end of four or five days, they are stapled in each student's booklet.

Preparation Time

Four or five days.

SNOWFLAKES

Description of Booklet

This booklet has no cover. It consists of 9" x 12" sheets of colored construction paper that are stapled together when completed by students.

Learning Skills Involved

Eye-hand coordination.

Method of Preparation

Large cardboard circle patterns are cut out by the teacher for students to trace on typing paper. Students are given typing paper, cardboard patterns, pencils, scissors, glue and 9" x 12" colored construction paper.

The student traces a cardboard circle on typing paper and then cuts it out. It is then folded and cut out to become a snowflake. (Instructions are given on the following page.) After cutting out the snowflake, it is glued on a sheet of colored construction paper.

Students who make more than one snowflake may have their papers stapled together to form a booklet.

Preparation Time

One day.

How to Make a Snowflake

1. A large cardboard circle is used as a pattern. It is traced on typing paper, then cut out.

2. The paper circle is folded in half.

3. It is then folded into thirds so it resembles a triangle with the short end rounded.

4. An irregularly shaped V is cut out of the rounded edge.

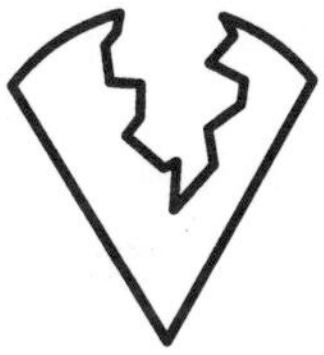

5. Smaller shapes are snipped out of the sides.

6. The snowflake is unfolded and glued on to a sheet of colored construction paper.

MY STATE

Description of Booklet

This booklet consists of a 12" x 18" cover of colored construction paper that is folded in half. Several prepared inner pages of maps are stapled inside the cover for students to complete.

Learning Skills Involved

Eye-hand coordination and visual discrimination.

Method of Preparation

Booklets are prepared in advance by the teacher. Each page contains either a map of the United States or a map of the state. Children will follow directions on each page. For example:

Page 1 Shows a map of the United States. Students are asked to color their state. They may be asked to put names on other states such as adjoining states.

Page 2 Shows a map of the state. Children are asked to write in the name of the city in which they live as well as the capital of the state.

Page 3 Shows a map of the state. Children are asked to draw the principal rivers in the state.

Page 4 Shows a map of the state. Children are asked to mark vacation spots and historical sites.

Preparation Time

Three or four days.

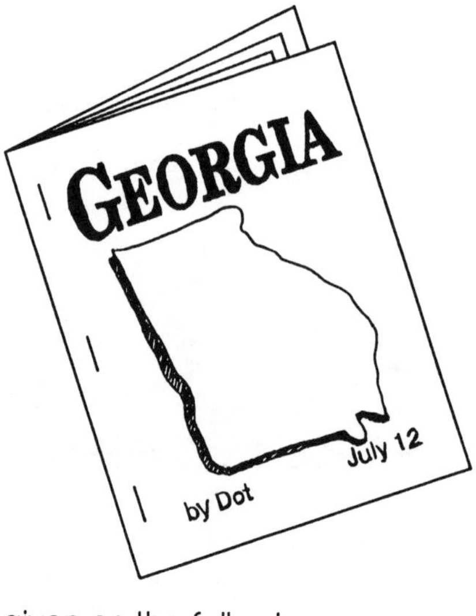

A map of the United States is given on the following page.

COLOR YOUR STATE

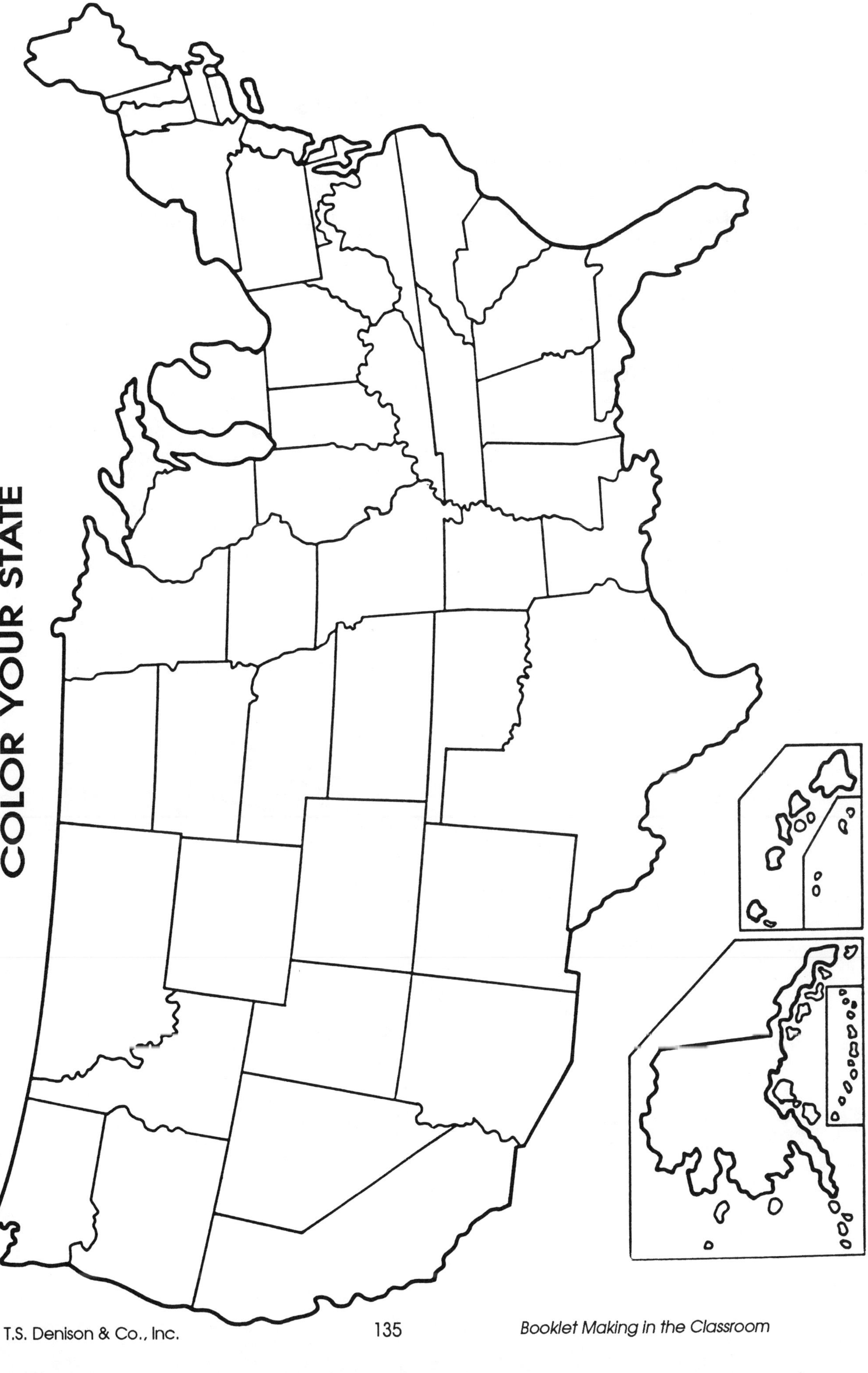

ISLANDS

Description of Booklet (Group Booklet)

This booklet consists of a cover of two 12" x 18" pieces of oak tag or colored posterboard. Inner pages of manila paper are inserted after all students have completed their work.

Learning Skills Involved

Observation and collecting information.

Method of Preparation

Maps and globes are made available to students. Each child selects one island and draws it on his/her paper. Pencils and crayons should be placed on the work table along with 12" x 18" manila paper.

Two or three facts about each island are printed on the page. Facts such as climate, location, industry, population, major cities, language spoken, etc., are examples of what can be used. Class discussion should precede all work.

Preparation Time

Two days. One day for selecting and drawing the island and one day for gathering information about the island.

EARTH

(Fire, Water or Air)

Description of Booklet

This booklet consists of a 12" x 18" cover of colored construction paper that is folded in half. Several inner pages of typing paper are stapled inside the cover. Booklets are prepared in advance by the teacher.

Learning Skills Involved

Observation, imagination, discrimination and summarization.

Method of Preparation

In ancient times, it was believed that there were four elements: earth, air, water and fire. Each child selects one of these subjects to investigate. Reference books and visits to the school library are suggested. After completion, students should share their findings with the class.

Class discussion should precede this project. The following ideas for each element can be explored:

Water: When helpful/when harmful, pollution causes, various sources of water.

Earth: Uses, types of earth around the school, around your home, around the world, pollution causes.

Fire: When helpful/when harmful, different causes of fire, describe a fireman's job.

Air: Kinds of winds, when helpful/when harmful, pollution causes.

Preparation Time

Two or three days.

MY BOOK OF CONTINENTS

Description of Booklet

This booklet consists of a 12" x 18" cover of colored construction paper that is folded in half. Seven pre-printed worksheets of continents are stapled inside the cover.

Learning Skills Involved

Observation and eye-hand coordination.

Method of Preparation

Booklets are constructed in advance by the teacher. Students are given pencils and crayons. Sample pictures of labeled continents are placed on the work table.

Children color the continents and then copy the names of the continents in the empty boxes.

For older students, rivers and mountain ranges may also be indicated.

Preparation Time

From three to five days.

Pattern for North America

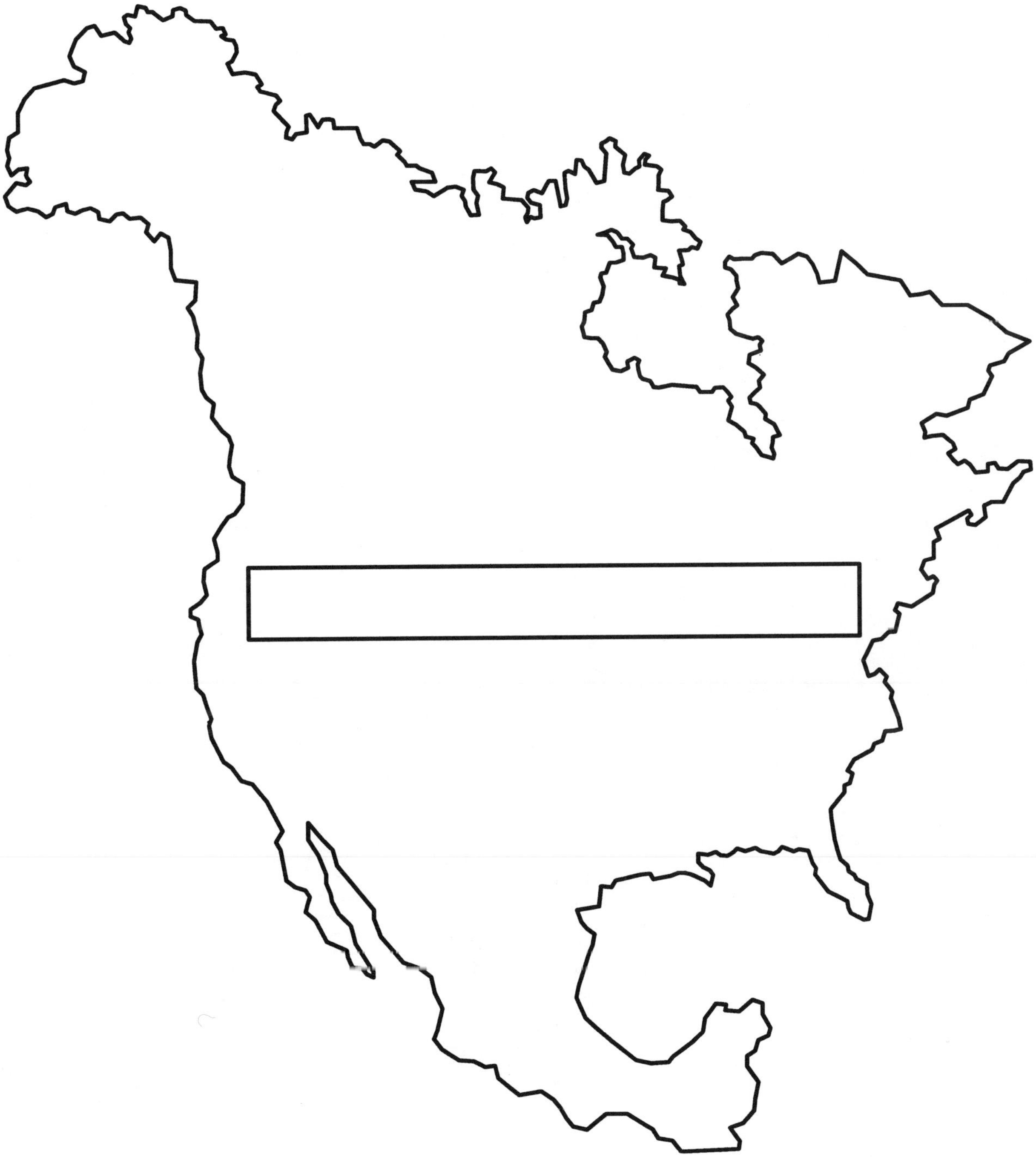

Pattern for South America

Pattern for Europe

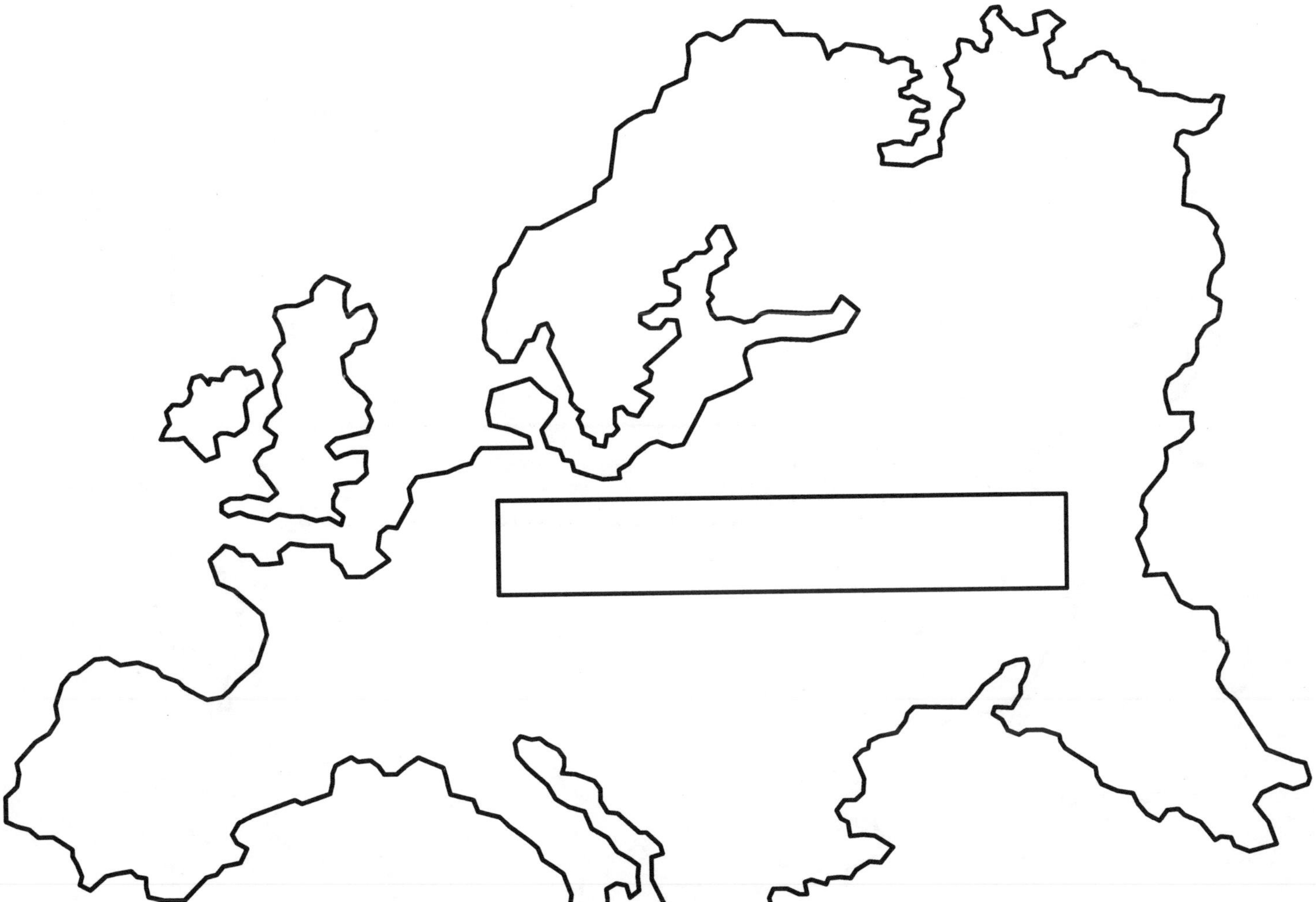

Pattern for Asia

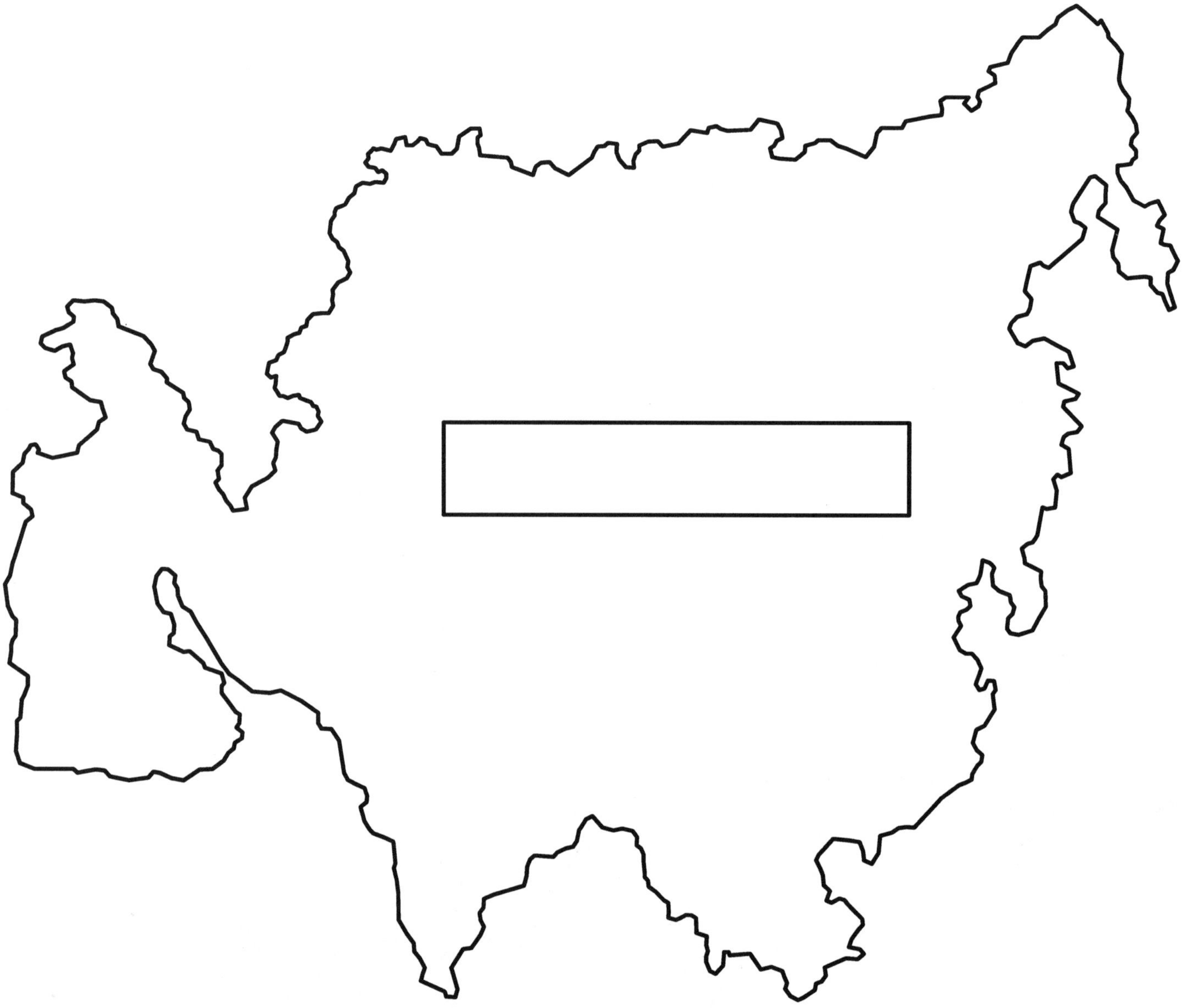

Pattern for Africa

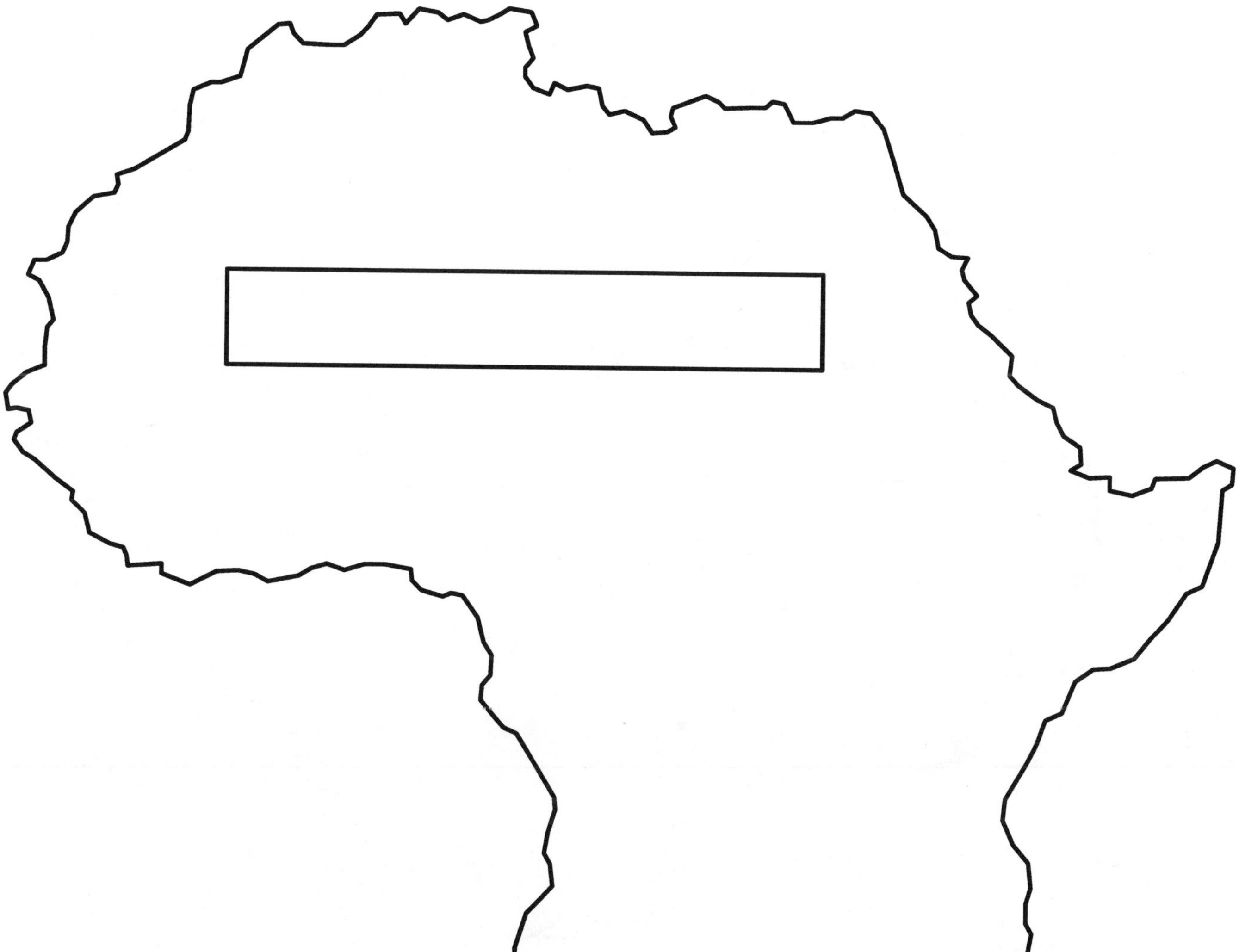

Pattern for Australia

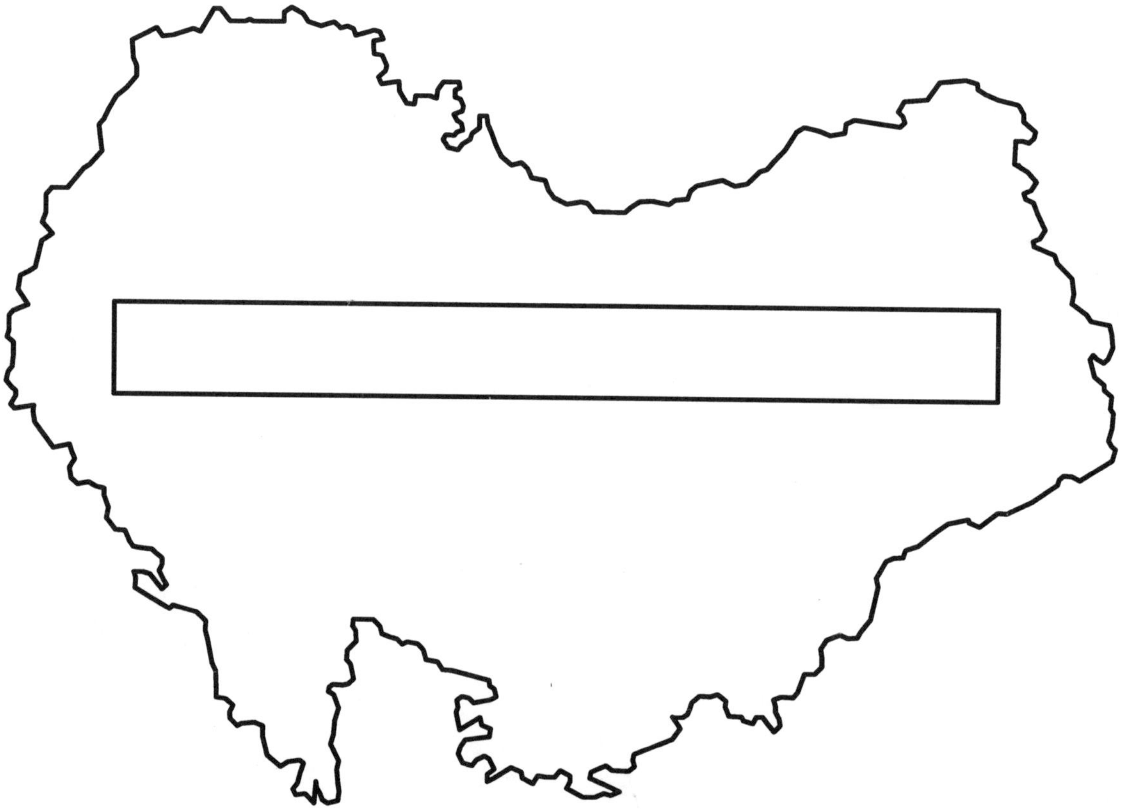

Pattern for Antartica

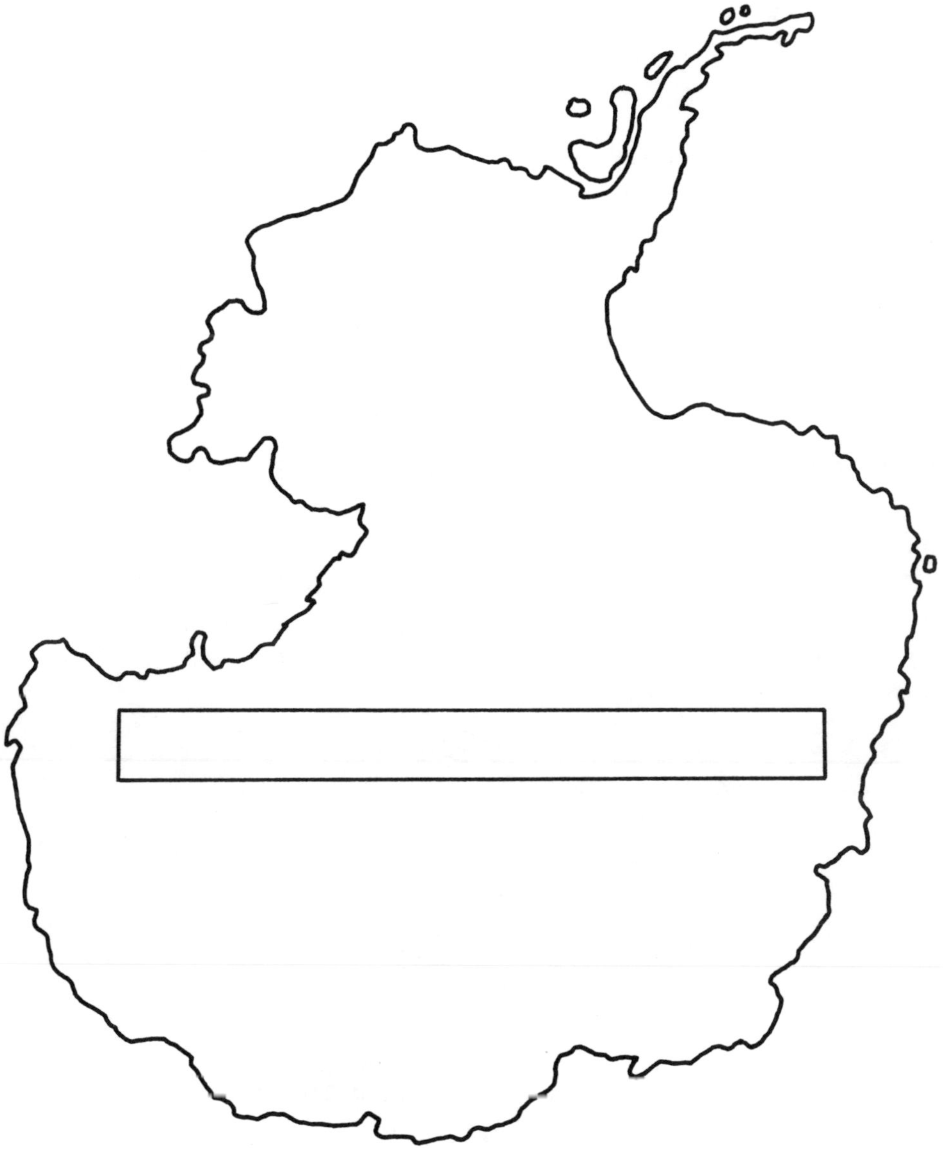

MY BOOK OF PLANETS

Description of Booklet

This booklet consists of a 12" x 18" cover of colored construction paper that is folded in half. Several inner pages containing drawings of planets by students are stapled inside the cover.

Learning Skills Involved

Eye-hand coordination and subject discrimination.

Method of Preparation

Nine various sized posterboard patterns of circles are prepared in advance by the teacher. Each circle represents a planet with its name clearly printed on it.

Students are given pencils to trace circles and colored markers to color planets and write in the names. They may draw one or more planets on each page.

Students should be encouraged to draw all nine planets.

Preparation Time

Two days for younger students. Four to five days for primary grade students. A single descriptive sentence should be written with each planet such as: "Pluto is the planet that is the farthest from the sun," or "Jupiter is the largest planet," etc.

Symbols for planetary objects are found on the next page.

Symbols of the Sun, Moon and Planets

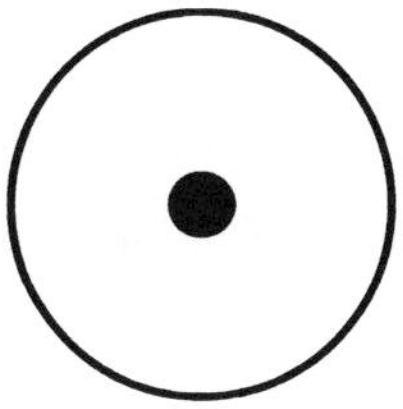

SUN

MOON

MERCURY

VENUS

EARTH

MARS

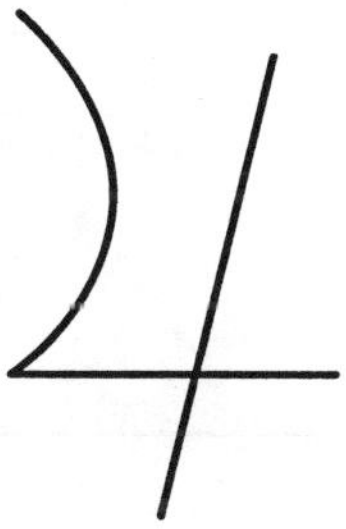

JUPITER

SATURN

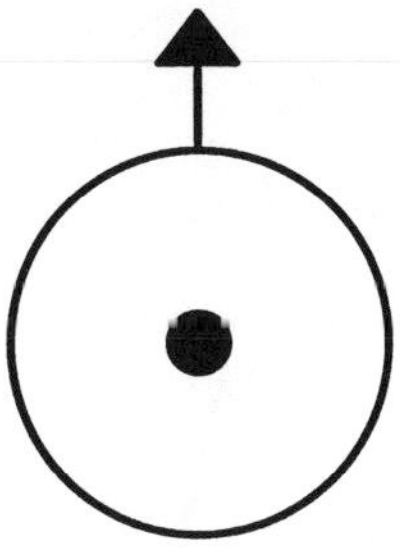

URANUS

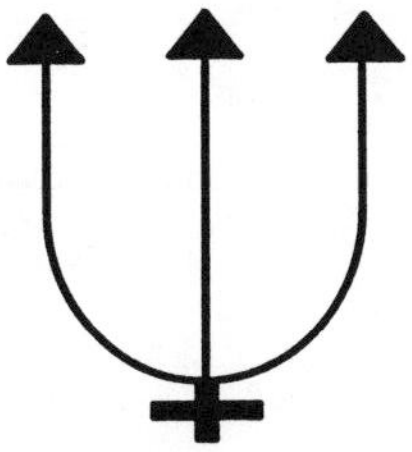

NEPTUNE

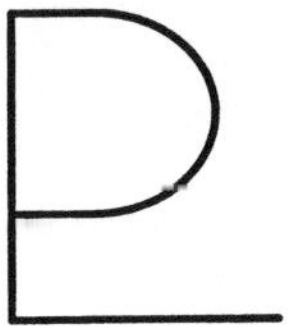

PLUTO

Students may draw these symbols on the matching planets in their booklets.

TRAVEL IN SPACE

Description of Booklet

This booklet consists of a 12" x 18" cover of colored construction paper that is folded in half. Four or five pages of typing paper are stapled inside the cover.

Learning Skills Involved

Imagination and observation.

Method of Preparation

Booklets are prepared in advance by the teacher. After each day's class discussion, children may be asked to draw one of the following:

1. draw an astronaut.
2. draw an astronaut floating in space (either in or out of the spacecraft).
3. draw a space craft.
4. draw a space craft landing on a planet and name the planet.
5. draw a space station.

Preparation Time

Four or five days.

CONSTELLATIONS

Description of Booklet

This booklet consists of a 12" x 18" cover of colored construction paper that is folded in half. Inner pages are stapled inside the cover.

Learning Skills Involved

Visual discrimination and eye-hand coordination.

Method of Preparation

Constellation papers are prepared in advance by the teacher. Typing paper or tracing paper is placed on the work table. Children copy a constellation and use crayons to trace the outline of the constellation. Stars are emphasized using Q-tips, glue and glitter.

Preparation Time

One or two days, depending how many constellations students decide to draw.

Constellation patterns are given on the following pages. They may be enlarged.

CONSTELLATION PATTERNS

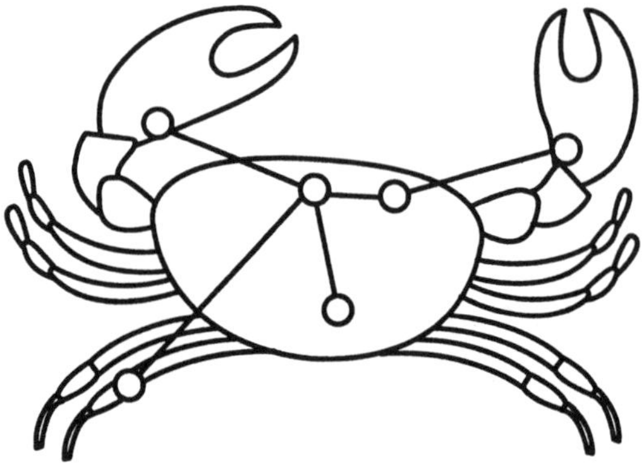

CANCER, the Crab

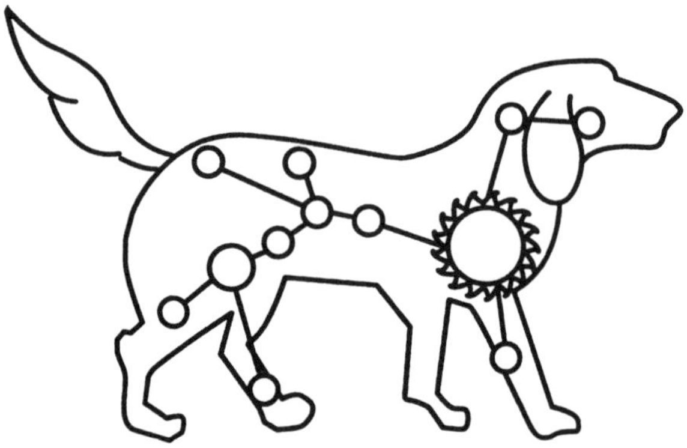

CANIS MAJOR, showing Sirius the Dog Star, the brightest star in the heavens

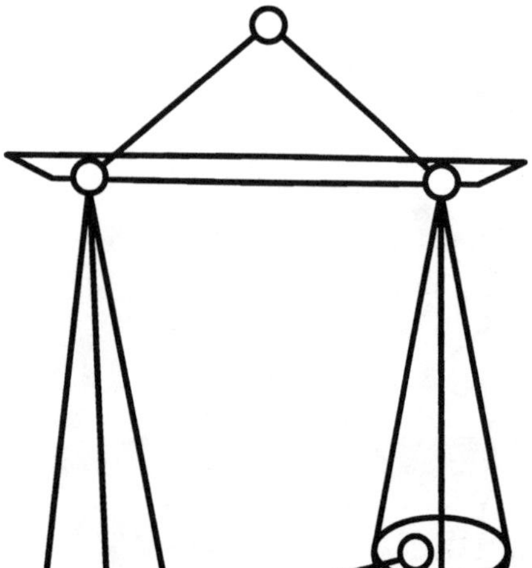

LIBRA, a pair of Scales

ARIES, the Ram

CONSTELLATION PATTERNS

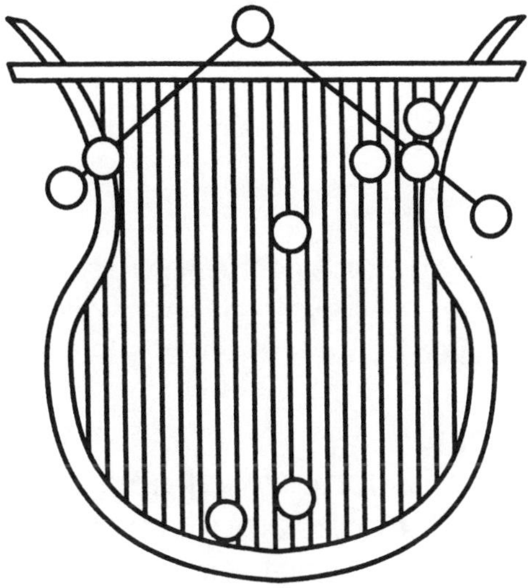

LYRA, the Harp

CAPRICORNIS, half fish–half goat

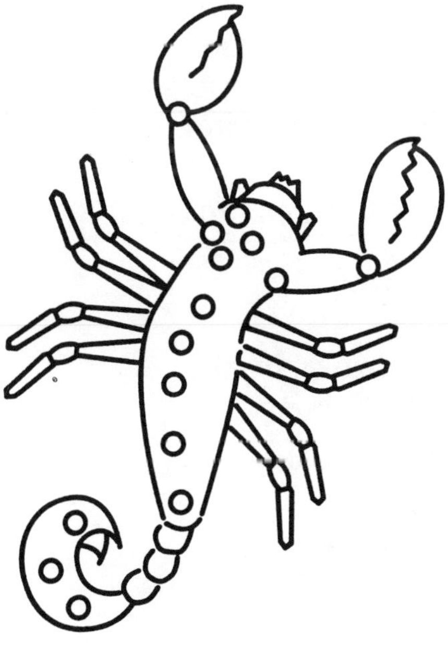

SCORPIUS, the Scorpion

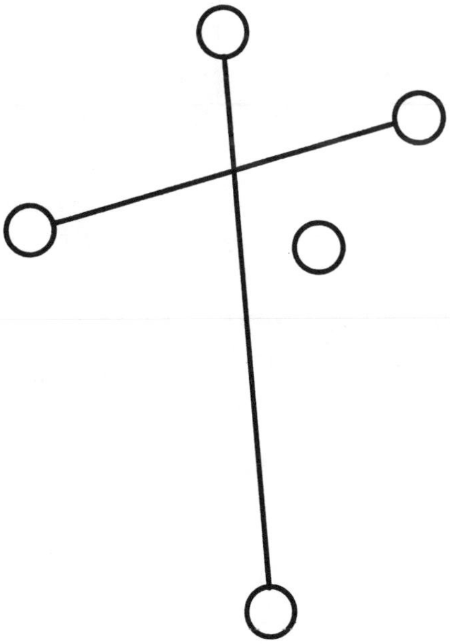

SOUTHERN CROSS

THREE SCIENCE EXPERIMENTS – #1

Description of Booklet

This booklet consists of a 12" x 18" cover of colored construction paper that is folded in half. Three duplicated pages are stapled to the inner back cover. Booklets are prepared in advance by the teacher.

Learning Skills Involved

Observation, classification and summarization.

Method of Preparation

After a class demonstration, the teacher sets up one of the three science experiments. It should be maintained for several days in order to allow all students to use it. The three experiments are:

1. Things that float/things that do not float
   Needed: A pan of water and several items such as a nail, coin, pencil, pin-pong ball, block of wood, chalk, empty spool, Ivory soap, sponge, key, knife, metal bracelet, etc.

2. Things attracted to a magnet/things not attracted to a magnet
   Needed: Magnets and items such as a nail, coin, cotton balls, thumb tacks, gem clips, eraser, chalk, pencil, key, etc.

3. Things that absorb water/things that do not absorb water
   Needed: Eye-dropper, a bowl of water and several items such as wax paper, newsprint, paper towel, toilet paper, tissue, fry-pan, aluminum foil, sponge, plastic wrap, mirror, etc.

Children work on each experiment independently or with a partner and record observations on their pages.

Preparation Time

Approximately three weeks; one week per experiment.

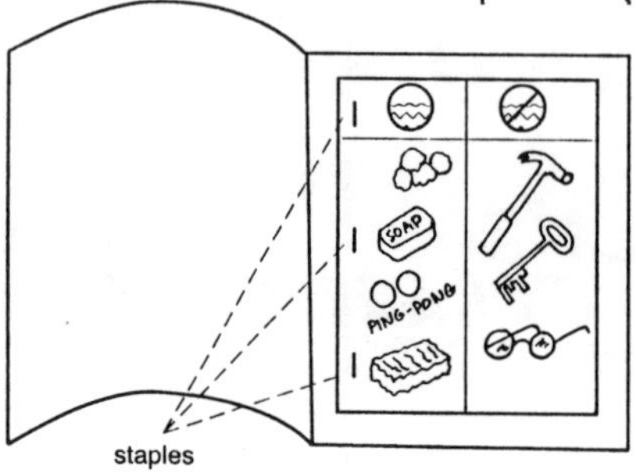

Headings for the three duplicated pages are given on the following page.

1. Cut out each strip and attach it to the top of each page befores duplicating.
2. Extend the center line to the bottom of each page.

THREE SCIENCE EXPERIMENTS – #2

Description of Booklet

This booklet consists of a 12" x 18" cover of colored construction paper that is folded in half. Three duplicated pages are stapled to the inner back cover. Booklets are prepared in advance by the teacher.

Learning Skills Involved

Observation, classification and summarization.

Method of Preparation

After a class demonstration, the teacher sets up one of the three science experiments. It should be maintained for several days in order to allow all students to use it. The three experiments are:

1. Rough/Smooth
   Needed: Items such as rocks, ping-pong balls, burlap, a mirror, a piece of aluminum foil, sandpaper, etc.

2. Hard/Soft
   Needed: Items such as cotton balls, a wooden block, a pencil, a key, a stuffed animal, pillow, play-dough, ruler, etc.

3. Heavy/Light
   Needed: Items such as an flashlight, hammer, sponge, cotton balls, brick, styrofoam peanuts, plastic ruler, etc.

Children handle the items and record their findings on inner pages (see illustration below) by writing the names of each object or drawing them in the proper column.

Preparation Time

Approximately three weeks; one week per experiment.

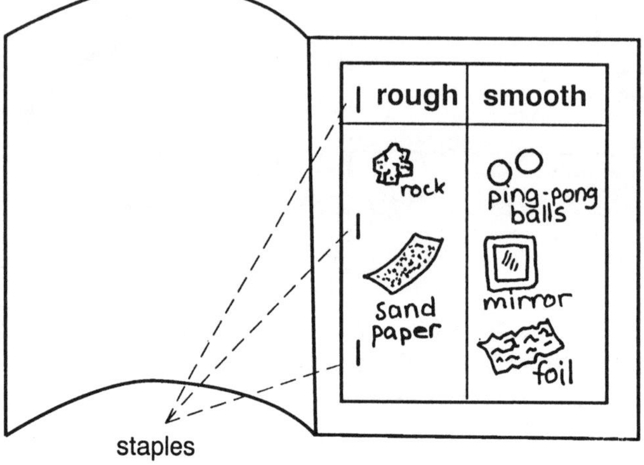

Chapter 5

SOCIAL STUDIES

The activities in this chapter are planned to enhance the following subjects:

- self-concept
- communication
- transportation
- occupations
- sports and hobbies
- economics
- rules and laws
- government
- history
- holidays
- religion

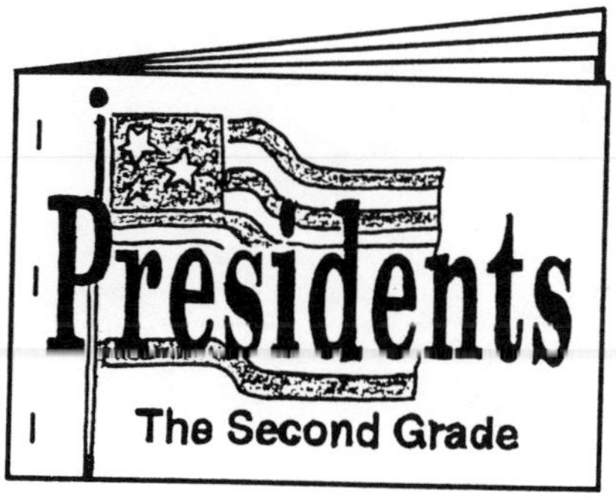

ALL ABOUT ME

Description of Booklet

This booklet consists of a 12" x 18" cover of colored construction paper that is folded in half. Four or five inner pages of typing paper are stapled inside the back cover. Booklets are prepared in advance by the teacher.

Learning Skills Involved

Eye-hand coordination and visual discrimination.

Method of Preparation

Each day the class discusses one aspect about a child that will contribute to a positive self-image. Pencils, crayons and/or felt-tipped markers are provided. The following five ideas are suggested:

Day 1	self-portrait
Day 2	child's handprint
Day 3	my home
Day 4	my pet (or a pet I would like to have)
Day 5	I like to . . . (favorite activity)

Preparation Time

Five days.

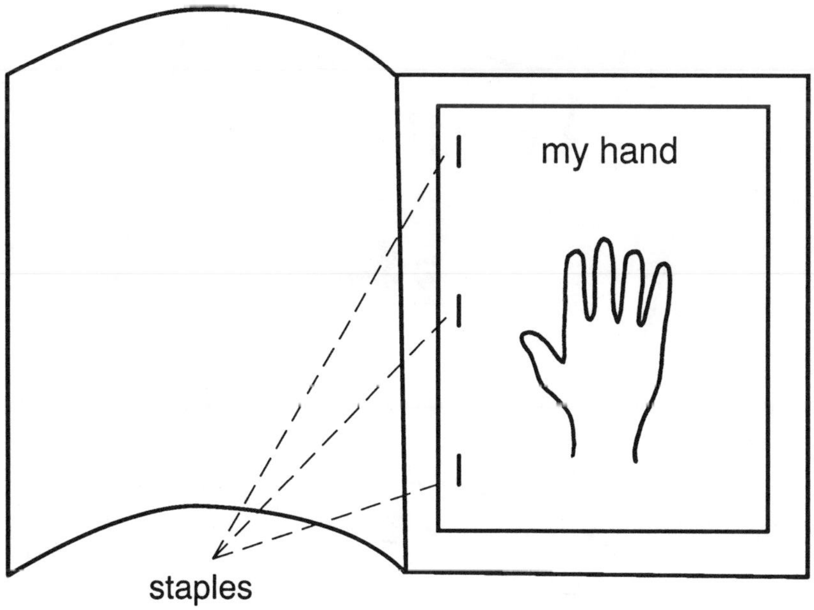

FAMILY ALBUM

Description of Booklet

This booklet consists of a 12" x 18" cover of colored construction paper that is folded in half. "Framed" inner pages are stapled inside the cover.

Learning Skills Involved

Imagination.

Method of Preparation

Booklets are prepared in advance by the teacher. A frame is drawn on an inner page before it is duplicated. Approximately four pages per booklet may be used.

Children are asked to draw and color family portraits – mother, father, sisters, brothers, grandparents, aunts, uncles, cousins, etc. All efforts should be accepted. Teachers or students may label pictures.

Preparation

Two or three days.

A sample "framed" page is given on the following page.

FAMILY ALBUM

Name____________________________________

FEELINGS

Description of Booklet

This booklet consists of a 12" x 18" cover of colored construction paper that is folded in half. Three or more inner pages of typing paper are stapled inside the cover. Booklets are prepared by the teacher.

Learning Skills Involved

Eye-hand coordination and reaching conclusions.

Method of Preparation

Each day the class discusses an emotion such as: happy, sad, angry, afraid, shy, etc. Students then illustrate this aspect in personal terms in their booklets.

Later in the day, the teacher will display booklets and lead a discussion. All attempts should be accepted and praised. The teacher will decide if names may be omitted when discussing the feeling.

Pencils, crayons and/or felt-tipped markers should be provided.

Preparation Time

Three or more days.

MY BIRTHDAY BOOK

Description of Booklet

This booklet consists of a 12" x 18" cover of colored construction paper that is folded in half. Inner pages of typing paper are stapled inside the cover. Booklets are prepared in advance by the teacher.

Learning Skills Involved

Eye-hand coordination and imagination.

Method of Preparation

Each day the class discusses one topic related to birthday celebrations. Encourage the children to share their own special birthday traditions. Then the students illustrate a unique detail of their birthday in their booklets. Topics might include:

1. my birthday cake
2. presents I would like to get
3. children singing "Happy Birthday" to me
4. me opening my presents

Pencils, crayons and/or felt-tipped markers are provided.

Preparation Time

Four days.

GUESS WHO?

Description of Booklet (Group Booklet)

This booklet consists of two 12" x 18" cover pages of oak tag or posterboard. Inner pages are manila paper using front and back sides, one side to a child.

Learning Skills Involved

Observation.

Method of Preparation

Each student is asked to bring a baby picture to school. The photograph is glued on his/her page. Clues about the students' appearance and likes and dislikes are written on each page by the teacher.

Preparation Time

None. Students take turns looking through the booklet helping to guess the identity of each student by the clues and the photograph.

TELEPHONE BOOKLETS

Children are given blank booklets in the shape of a telephone. They collect the names and telephone numbers of their classmates by having each sign on a line of the booklet.

Another method is to list each child's name and phone number on a chart and children copy them in their own telephone booklets.

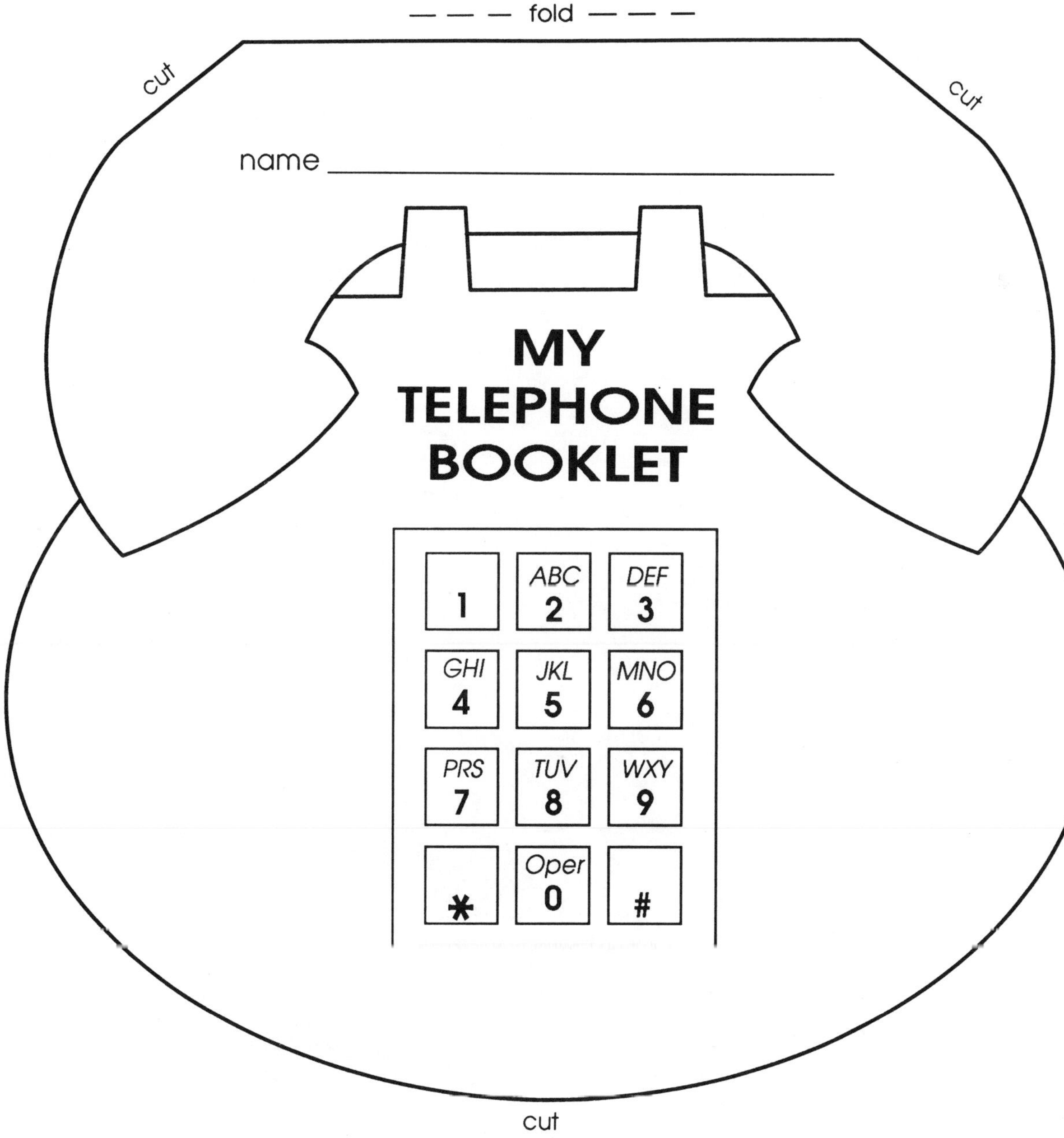

The above telephone booklet cover is traced on a folded piece of 12" x 18" construction paper. Various colors may be used for different pupils. Provide each child with 3-4 pages of first grade writing paper cut to the shape of the telephone pattern. Insert the pages and staple inside.

Separate dials may be given to the children to paste on the booklets.

COMMUNICATION

Description of Booklet

This booklet consists of a 12" x 18" cover of colored construction paper that is folded in half. Four or five inner pages of typing paper are stapled inside the cover. Booklets are made in advance by the teacher.

Learning Skills Involved

Eye-hand coordination, observation and imagination.

Method of Preparation

The class will discuss various means of communication such as: television, radio, telephones, newspapers, magazines, books, telegrams, letters, tapes, etc.

Students will then draw and title as many representative forms as they desire on their papers. Pencils, crayons and/or felt-tipped markers should be provided.

Preparation Time

Two or three days.

BOOK OF STAMPS

Description of Booklet

This booklet consists of a duplicated worksheet containing an envelope and two inner pages. When cut out, the cover and pages are stapled together.

Learning Skills Involved

Observation.

Method of Preparation

At the beginning the of the school year, students and teacher bring cancelled stamps to school. The stamps should be kept in a separate box.

When the box is full, usually in the spring, the project may begin. Each student is given a pre-printed page showing an envelope and two blank rectangles to be used as inner pages. The student cuts them out and staples them together to form a booklet with the envelope on top. The student fills in the blank spaces on the envelope.

Glue is furnished. Children search through the stamps and glue them in their booklet, trying not to duplicate stamp designs.

Preparation Time

One day.

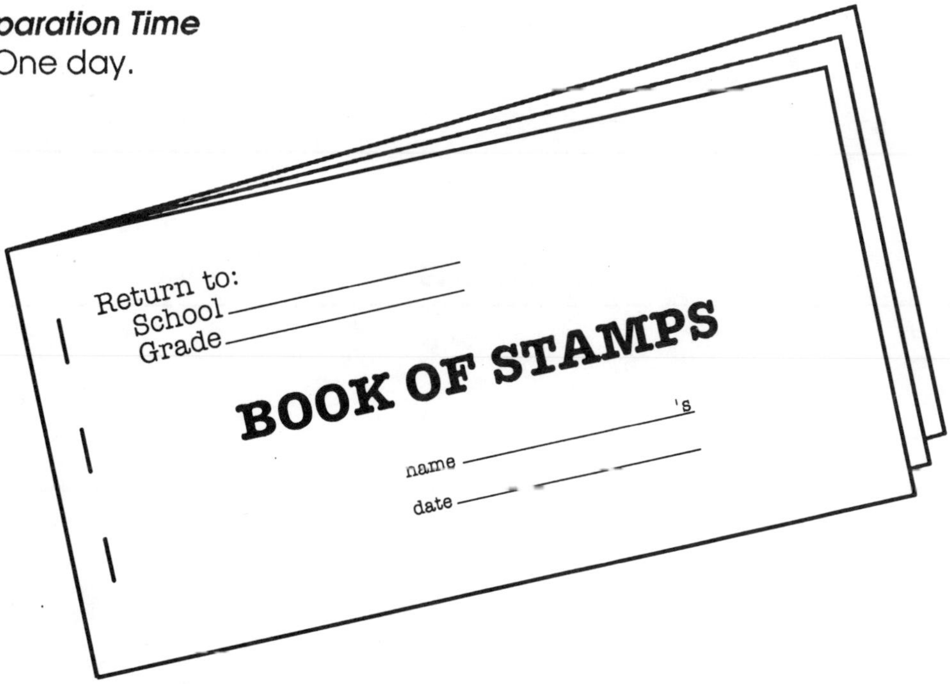

A pattern for an envelope and inner pages is given on the following page.

Return To:
School ______________________

Grade ______________________

29¢
USA

name ______________________'s

BOOK OF STAMPS

date ______________________

- -
cut here

STAMPS

- -
cut here

STAMPS

TRANSPORTATION

Description of Booklet

This booklet consists of a 12" x 18" cover of colored construction paper that is folded in half. Four or five inner pages of typing paper are stapled inside the cover. Booklets are made in advance by the teacher.

Learning Skills Involved

Eye-hand coordination, observation, and imagination.

Method of Preparation

The class will discuss various modes of transportation such as: cars, airplanes, boats, busses, spacecrafts, sleds, skis, skates, trucks, trains, trailers, horses, camels, donkeys, elephants, etc.

Students will then draw and title as many transportation forms as they desire on their papers. Pencils, crayons, and/or felt-tipped markers should be provided.

Preparation Time

Two or three days.

MY BOOK OF WHEELS

Description of Booklet

This booklet consists of a 9" x 12" cover of colored construction paper that is folded in half. Inner pages of typing paper cut in half are stapled inside to the back cover. Booklets are prepared in advance by the teacher.

Learning Skills involved

Imagination and innovation.

Method of Preparation

The class discusses vehicles and any other machine that uses wheels to function. Pencils, crayons and/or felt-tipped markers are provided.

Students draw a different means of transportation on each page. Pages should be labeled. The following "wheels" may be discussed:

car	scooter	mobile home	skate-board
truck	roller blades	tractor	motorcycle
bus	tricycle	bicycle	water wheel
train	wheel chair	wagon	ferris wheel

Preparation Time

Two or three days.

SAFETY SIGNS

Description of Booklet

This booklet consists of a 9" x 12" cover of colored construction paper that is folded in half. Five or six inner pages of typing paper cut in half are stapled to the inside back cover. Booklets are prepared in advance by the teacher.

Learning Skills Involved

Eye-hand coordination and observation.

Method of Preparation

Various patterns of safety signs constructed from properly colored posterboard are placed on the worktable. Children trace and color them.

Pencils, crayons and/or felt-tipped markers are provided.

Preparation Time

One or two days.

Sample patterns of safety signs are given on the following pages.

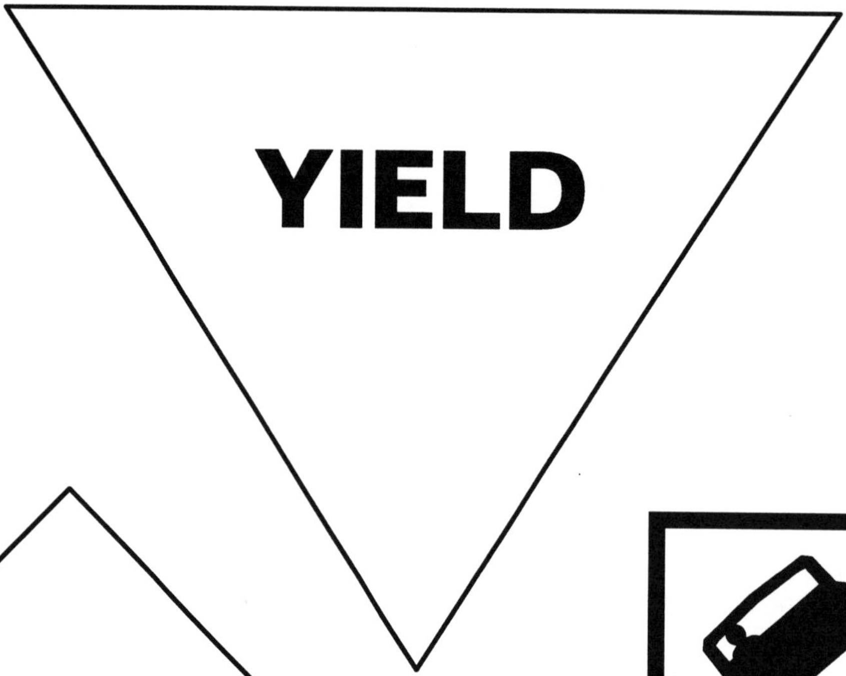
YIELD

SLOW

R R

STOP

ONE
WAY

GO! GO! GO!

Description of Booklet (Group Booklet)

This booklet consists of two 12" x 18" cover pages of oak tag or posterboard. Inner pages of manila paper are inserted upon completion by students.

Learning Skills Involved

Creativity and following directions.

Method of Preparation

After a class discussion, children are asked to illustrate an imaginary trip that they would like to take as well as their desired form of transportation.

Aside from the usual cars, busses, boats or trains; children might be encouraged to think of unusual methods of getting somewhere such as rockets, skis, balloons, camels, magic carpets, pogo sticks, broomsticks, etc. Imagination and whimsey can be part of the project.

12" x 18" manila paper, pencils, crayons and/or felt-tipped markers are provided. Drawings should be labeled either by students or by the teacher.

Preparation Time

One day.

I WANT TO BE A . . .

Description of Booklet

This booklet consists of a 12" x 18" cover of colored construction paper that is folded in half. Three or four inner pages of typing paper are stapled inside. Booklets are prepared in advance by the teacher.

Learning Skills Involved

Imagination and decision-making.

Method of Preparation

A class discussion is held prior to booklet work. Various occupations may be suggested. A few are:

carpenter	veterinarian	florist	preacher/priest/rabbi
doctor/nurse	chef	musician	fireman
fisherman	flight attendant	artist	cowboy
farmer	singer	teacher	jockey
policeman	dancer	astronaut	racer
pilot	athlete	TV announcer	actor

Each student selects an occupation and describes it in words, illustrations and/or cut-out pictures from magazines. Pencils, crayons and/or felt-tipped markers, scissors and paste are provided. The following questions may assist:

1. Where does this person work?
2. When does this person work?
3. What does this person wear?
4. How does this person train for the job?
5. What does this person do on the job?

Preparation Time

Three or four days.

INVENTORS

Description of Booklet (Group Booklet)

This booklet consists of two 12" x 18" cover pages of oak tag or posterboard. Inner pages of manila paper are inserted after completion by students.

Learning Skills Involved

Observation, evaluation and summarization.

Method of Preparation

A class discussion on inventors is held. Each student selects one inventor to research. Reference books and encyclopedias are made available along with library privileges.

Pencils, pens and/or felt-tipped markers are provided. Illustrations should be encouraged.

A sample list of inventors and their inventions is given on the following page. (Additional inventors may be added by teachers or students.)

Preparation Time

Approximately two days.

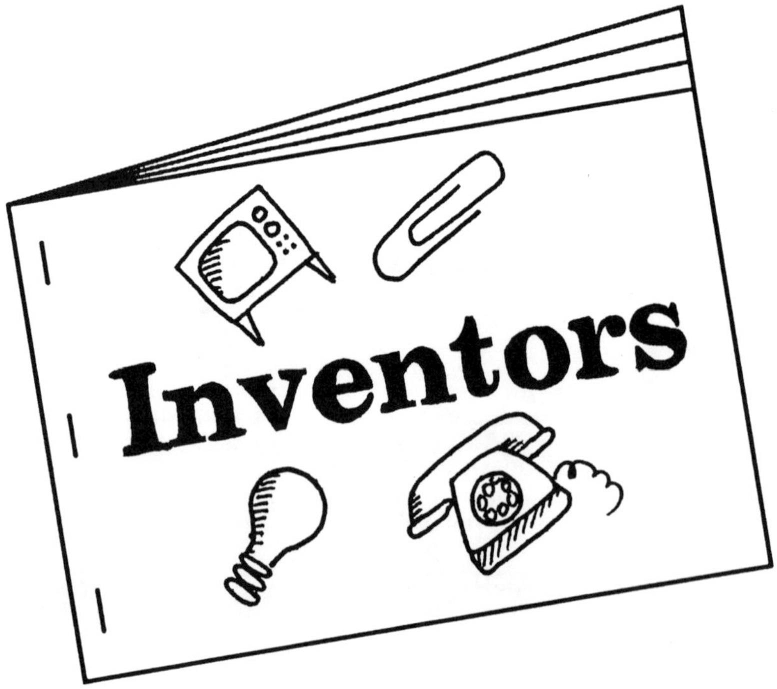

INVENTORS

Alexander Graham Bell	telephone (1876)
Walter Brattain John Bardeen William Shockley	transistor (1948)
Willis H. Carrier	air conditioner (1902)
Thomas Edison	incandescent light (1879) phonograph (1877)
Henry Ford	mass production (assembly line) of automobiles (1903)
Robert H. Goodard	liquid fuel rocket (1926)
Elias Howe	sewing machine (1846)
Walter Hunt	safety pin (1849)
W. L. Judson	zipper (1896)
John Loud	ball point pen (1888)
Theodore Maiman	laser (1960)
Guglielmo Marconi	radio (1896)
Samel F. B. Morse	telegraph (1837)
Elisha G. Otis	elevator (1853)
Wilhelm Roentgen	x-rays (1895)
Percy Spencer	microwave ovens (1946)
Eli Whitney	cotton gin (separtes cotton fiber and seed; 1793)
Wright Brothers	airplane (1903)
Vladimir K. Zworykin	television (1923)

THE OLYMPICS

Description of Booklet

This booklet consists of a 12" x 18" cover of colored construction paper that is folded in half. Several inner pages of typing paper are stapled inside the cover.

Learning Skills Involved

Eye-hand coordination, observation, and investigation.

Method of Preparation

It is suggested that this booklet be prepared during a year that the Olympics are held. Class discussions of the first Olympics will introduce the topic. Various sports are listed such as: swimming, diving, running, jumping, gymnastics, pole vaulting, ice skating, skiing, horseback riding, basketball, volley ball, soccer, etc.

Students may illustrate a different sport on each page.

Preparation Time

Three or four days.

HOT AIR BALLOONS

Description of Booklet (Class Booklet)

This booklet consists of a 12" x 18" cover page of colored construction paper or oak tag. Inner pages are duplicated worksheets of hot air balloons.

Learning Skills Involved

Eye-hand coordination and creativity.

Method of Preparation

Each child is given a duplicated drawing of a hot air balloon to color. Crayons and/or felt-tipped markers are placed on the work table. If desired, additional material may be made available for decorating balloon pictures such as glue, sequins, yarn, glitter, etc.

When completed, each child's balloon is stapled in the booklet. This booklet may be kept on the bookshelf along with library books.

Preparation Time

One or two days.

A sample of a hot air balloon drawing is given on the following page.

HOT AIR BALLOON

Name____

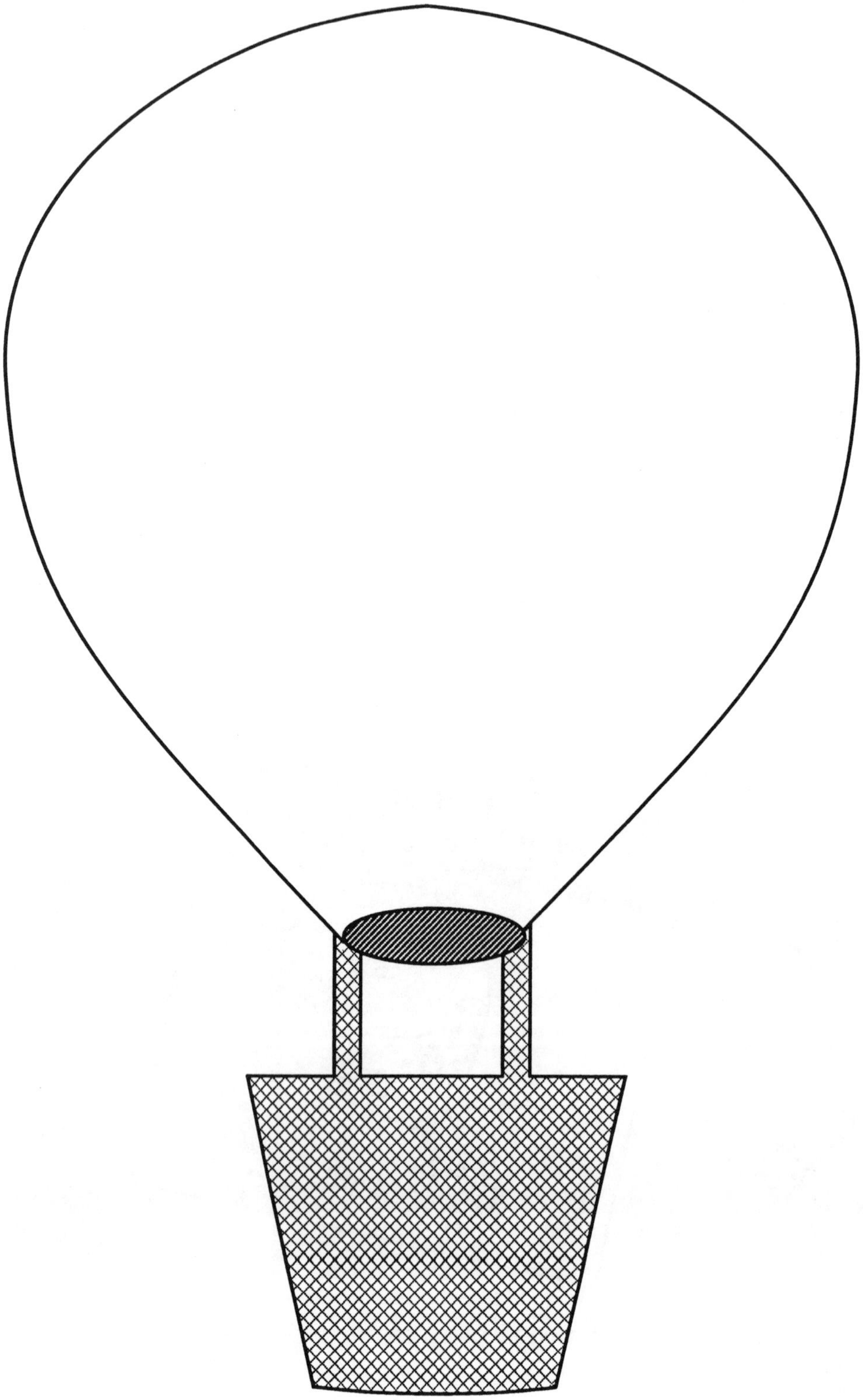

MY MONEY BOOK

Description of Booklet

This booklet consists of a 12" x 18" cover of colored construction paper that is folded in half. Several inner pages of typing paper will be stapled inside the cover after students have completed them.

Learning Skills Involved

Eye-hand coordination and recognition of coins.

Method of Preparation

Booklet covers are prepared in advance for each student. Several pennies (dimes, nickels, etc.) are placed in the center of the worktable. Typing paper and crayons are offered. Students place a penny under their paper and using the long side of a crayon, rub over it. Many pennies may be rubbed on one sheet, using a variety of colored crayons. Students should be asked to rub the front and back side of coins.

Preparation Time

One page per day of each coin offered.

SELLING A PRODUCT

Description of Booklet

This booklet consists of a 12" x 18" cover page of construction paper that is folded in half. Several pages of lined or typing paper are stapled inside the cover after students have completed them.

Learning Skills Involved

Creativity is emphasized.

Method of Preparation

After a class discussion, the teacher and class decide on a list of possible products to "sell" such as toys, sports, food, records, cereal, vacations, pets, etc. Means of selling are discussed, i.e., television, newspaper, radio or magazines. Ways of describing a product are investigated.

Children then draw the product and "sell" it.

Preparation Time

Depending on the number of products, approximately three days.

FIRST GRADE RULES

Description of Booklet (Group Booklet)

This book consists of two 12" x 18" cover pages of oak tag or posterboard. Inner pages of manila paper are to be inserted upon completion by students. The booklet will be held together by brads.

Learning Skills Involved

Imagination, observation, and reaching conclusions.

Method of Preparation

A class discussion concerning laws and rules will ensue prior to booklet work. Each student will then write one rule and illustrate it on his/her paper.

Pencils, crayons, and/or felt-tipped markers are provided.

Preparation Time

One day.

THE ELECTION OF *(DATE)*

Description of Booklet

This booklet consists of a 12" x 18" cover of colored construction paper that is folded in half. Inner pages of typing paper are stapled inside the cover.

Learning Skills Involved

Following directions, observing and summarizing.

Method of Preparation

Various candidates running in current elections are discussed and listed on a bulletin board. Students and teacher are asked to bring newspaper clippings and magazine photographs of candidates. These can be kept in a box or on the bulletin board.

Students describe each candidate on a separate page. The following information may be included:

1. age of the candidate
2. candidate's home state
3. candidate's political party
4. candidate's occupation
5. candidate's family
6. if possible, one point on the candidate's platform

A newspaper photo of the candidate (taken from the clippings or brought from home) is pasted on each page.

Preparation Time

Approximately one week.

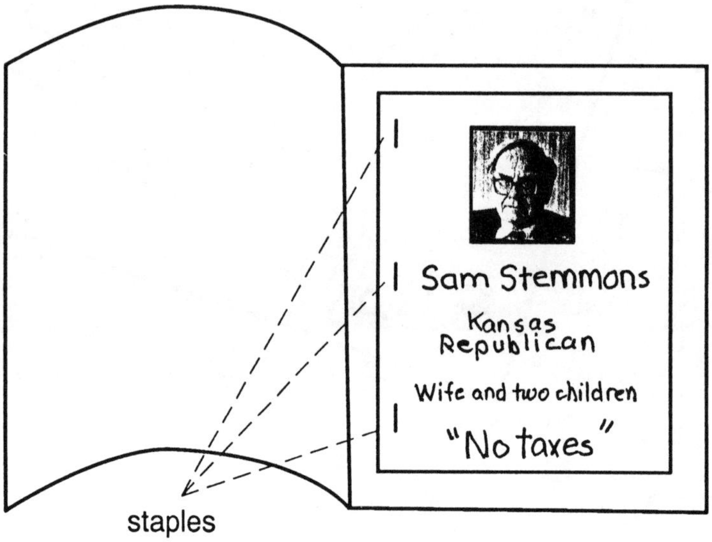

PRESIDENTS

Description of Booklet (Group Booklet)

This booklet consists of two 12" x 18" cover pages of oak tag or posterboard. Inner pages of manila paper are stapled in chronological order inside the cover by the teacher upon completion by students.

Learning Skills Involved

Observation and following directions.

Method of Preparation

Each student selects one president for investigation. The following facts should be included on papers:

birthdate and death date
years served as president
state in which he was born
occupation
political party to which he belonged
one important historical event that occurred during his term in office

Reference books, paper, pencils, and crayons should be provided. Students will write facts and, if possible, illustrate their papers.

Preparation Time

Two or three days.

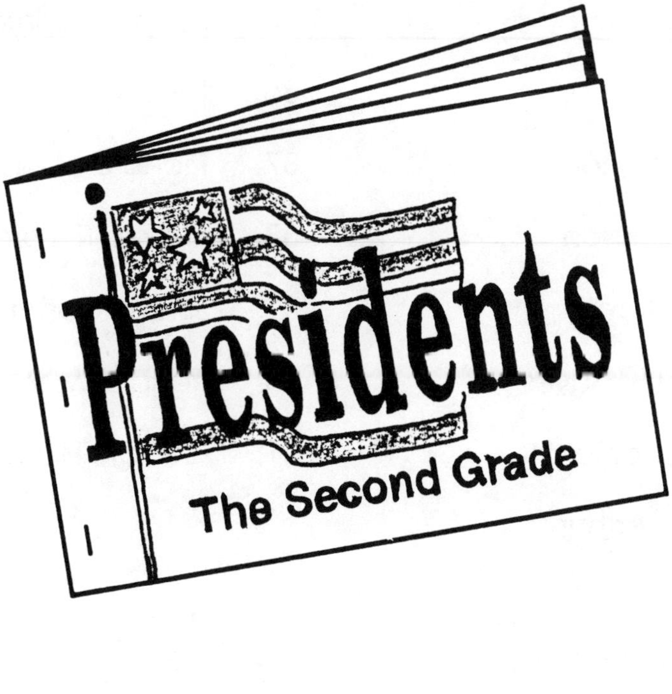

PRESIDENTS

1. George Washington
2. John Adams
3. Thomas Jefferson
4. James Madison
5. James Monroe
6. John Quincy Adams
7. Andrew Jackson
8. Martin Van Buren
9. William Henry Harrison
10. John Tyler
11. James K. Polk
12. Zachary Taylor
13. Millard Fillmore
14. Franklin Pierce
15. James Buchanan
16. Abraham Lincoln
17. Andrew Johnson
18. Ulysses S. Grant
19. Rutherford B. Hayes
20. James A. Garfield
21. Chester A. Arthur
22. Grover Cleveland
23. Benjamin Harrison
24. Grover Cleveland
25. William McKinley
26. Theodore Roosevelt
27. William Howard Taft
28. Woodrow Wilson
29. Warren G. Harding
30. Calvin Coolidge
31. Herbert C. Hoover
32. Franklin D. Roosevelt
33. Harry S. Truman
34. Dwight D. Eisenhower
35. John F. Kennedy
36. Lyndon B. Johnson
37. Richard M. Nixon
38. Gerald R. Ford
39. Jimmy Carter
40. Ronald W. Reagan
41. George H. Bush

EXPLORERS

Description of Booklet

This booklet consists of a 12" x 18" cover of colored construction paper that is folded in half. Three or four inner pages of typing paper are stapled inside the cover.

Learning Skills Involved

Observation, decision-making, and summarizing.

Method of Preparation

A class discussion about explorers is held. Where were their homelands? Why did they become explorers? What did they accomplish? A list of explorers – as suggested by students – may be posted on a bulletin board. Some famous explorers include: Marco Polo, Eric the Red, Vasco deGama, Hernando DeSoto, Hernando Cortez, Ferdinand Magellan, James Cook, Meriwether Lewis and William Clark, Sir Henry Stanley, Robert Peary, etc.

It is suggested that each student select one explorer to investigate. Reference books and biographies may be placed on the work table. Imaginative illustrations are to be encouraged.

Preparation Time

Two or three days.

HALLOWEEN

Description of Booklet

This booklet consists of a 12" x 18" cover of colored construction paper that is folded in half. Four or five inner pages of typing paper are stapled to the inside back cover. Booklets are prepared in advance by the teacher.

Learning Skills Involved

Eye-hand coordination and imagination.

Method of Preparation

A Halloween story is read to the class. Children will then illustrate the story on their pages. An alternative method may include children drawing one aspect of Halloween on each page. For example:

Page 1 Witches and Ghosts
Page 2 My Costume
Page 3 Trick or Treaters
Page 4 A Halloween Pumpkin

Preparation Time

One or two days.

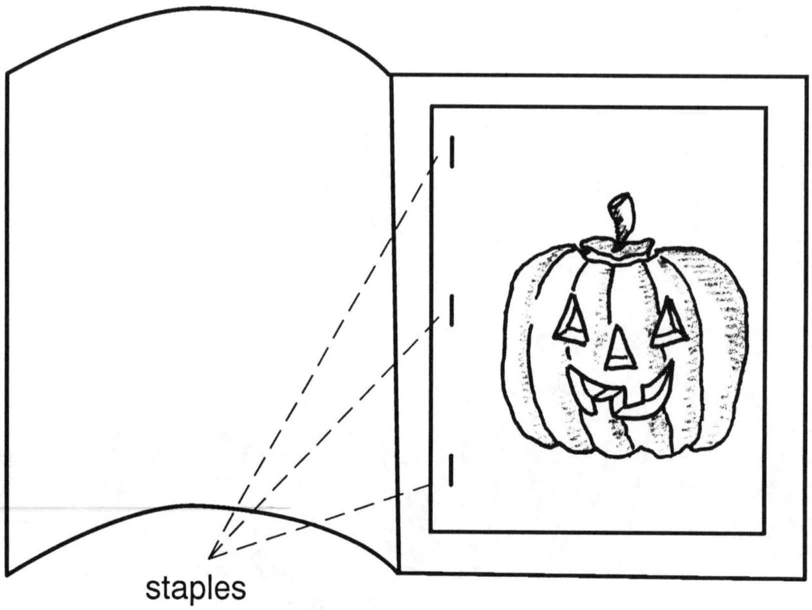

THANKSGIVING

Description of Booklet

This booklet consists of a 12" x 18" cover of colored construction paper that is folded in half. Three or four 9" x 12" manila pages are stapled inside the cover.

Learning Skills Involved

Eye-hand coordination, listening, and following directions.

Method of Preparation

Cardboard patterns for cornucopias are placed on the work table for students to trace, color, cut out, and paste on the cover of the booklet.

Each page in the booklet illustrates one aspect of Thanksgiving. The following suggestions may be used:

1. The Mayflower
2. Plymouth Rock
3. Pilgrims
4. Indians
5. The First Thanksgiving Feast

Students are offered pencils, crayons, paste, and scissors.

Preparation Time

Three or four days.

Cornucopia pattern is found on the following page.

Pattern for

THANKSGIVING CORNUCOPIA

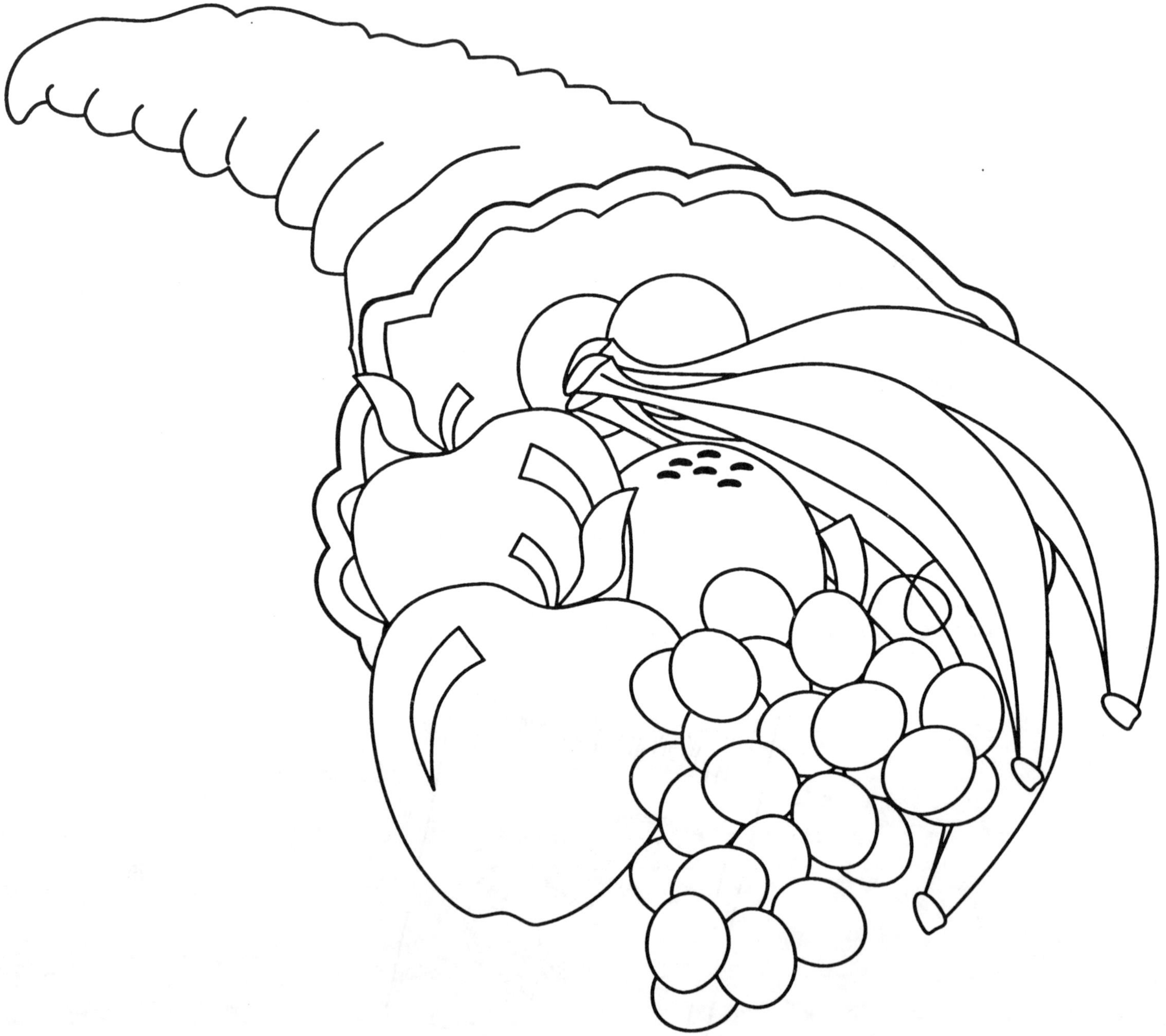

NATIVE AMERICANS

Description of Booklet

This booklet consists of art work and worksheets. It has a 12" x 18" cover of colored construction paper that is folded in half. Inner pages are stapled inside upon completion by students.

Learning Skills Involved

Eye-hand coordination, visual discrimination, and observation.

Method of Preparation

A teepee is traced, cut out, and decorated by students and pasted on the cover. Inner pages are:

Page 1 match numbers to feathers on a headdress
Page 2 match arrowheads
Page 3 weaving

Patterns, worksheets, and instructions are given on the following pages.

Preparation Time

Five days.

Indians of long ago and today are discussed, taking into consideration different tribes, customs, languages, homes, etc. This booklet is an excellent accompaniement to Thanksgiving for Pre-K and Kindergarten students.

Pattern for

TEEPEE

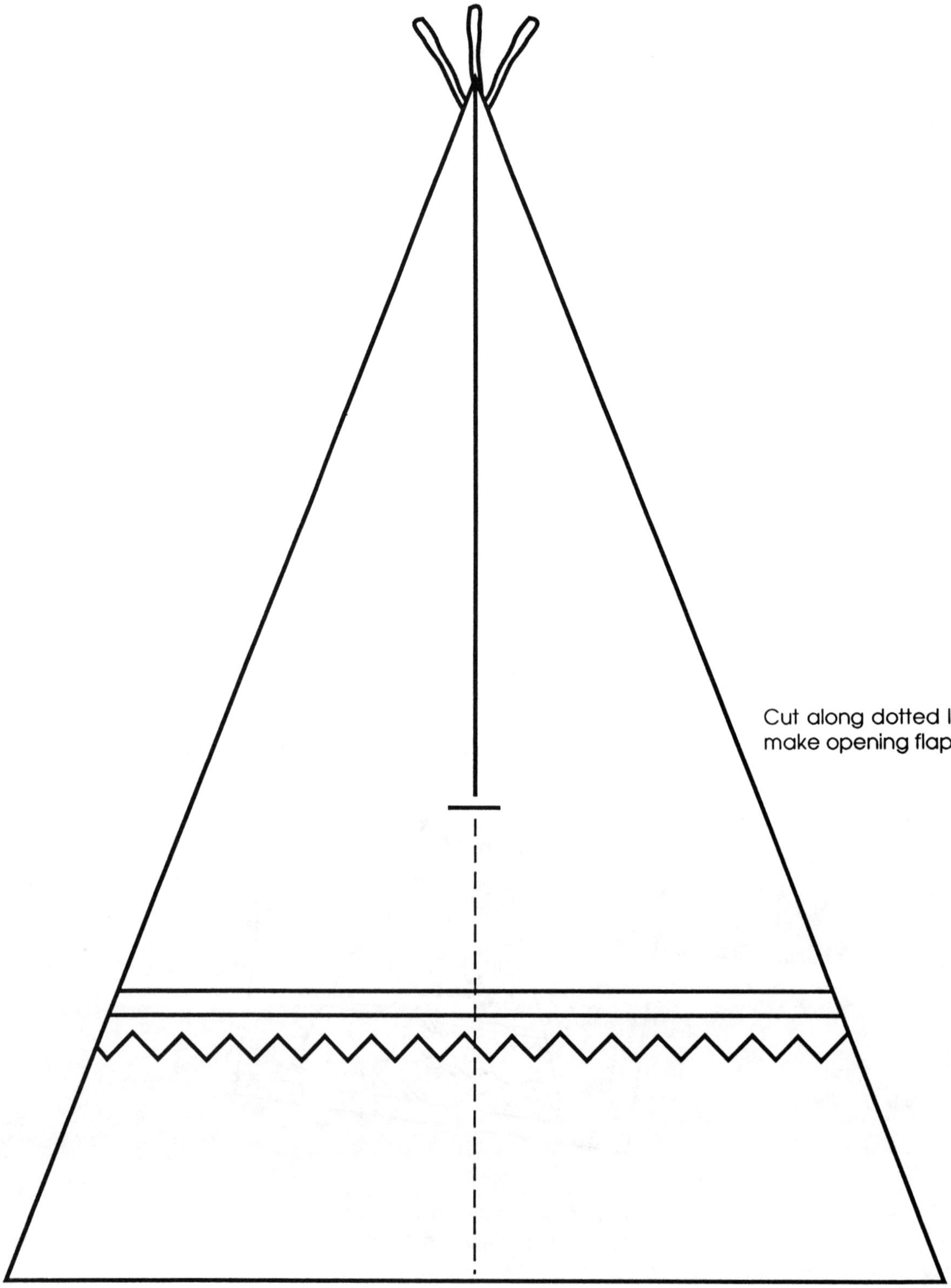

Cut along dotted line to make opening flap.

COUNTING WORKSHEET

Name____

4

2

5

3

1

ARROWHEAD MATCHING WORKSHEET

Name____

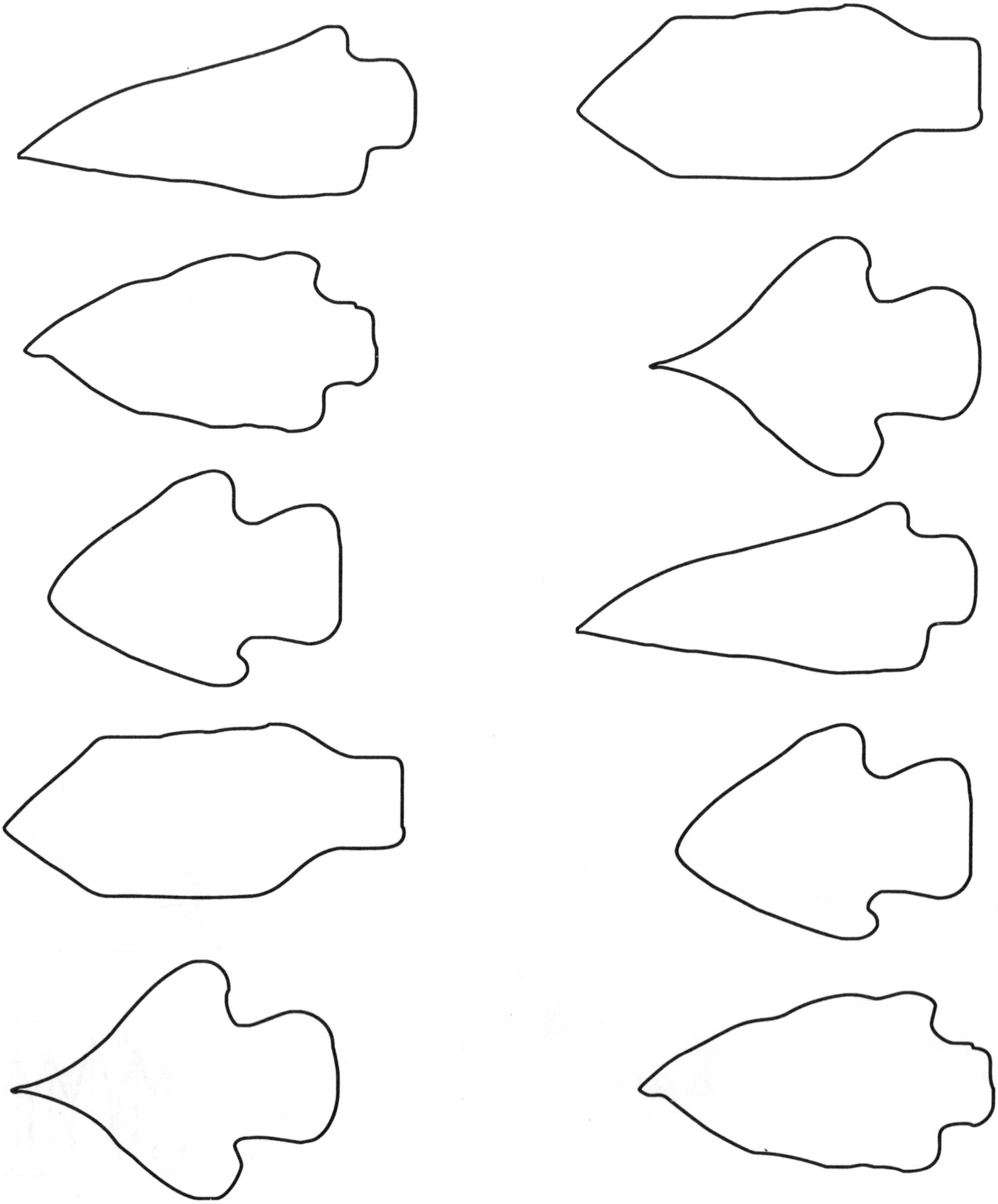

Instructions for

WEAVING PAPER

A folded piece of 9" x 12" colored (or manila) construction paper is used. The teacher draws straight guide lines from the fold to within 1 1/2 to 2 inches short of the cut edge.

Children cut these parallel slits on the guide lines – being careful not to cut to the edge.

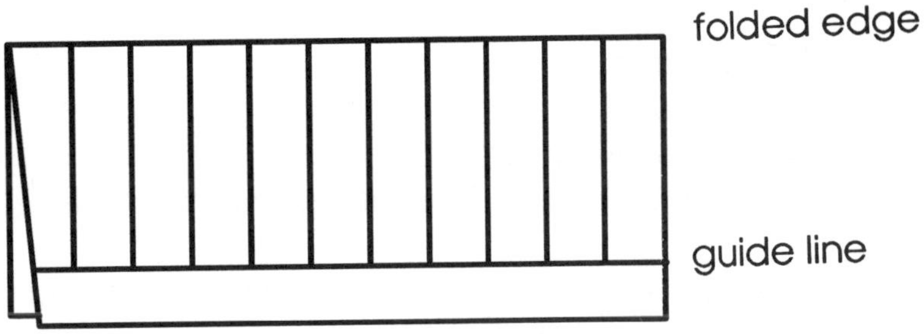

Using two or more colors, 12" strips of paper are provided for weaving. Students flatten their papers and weave in and out of the slits using one strip at a time. Strips are inserted alternately over and under the paper. Using two colors simplifies the task.

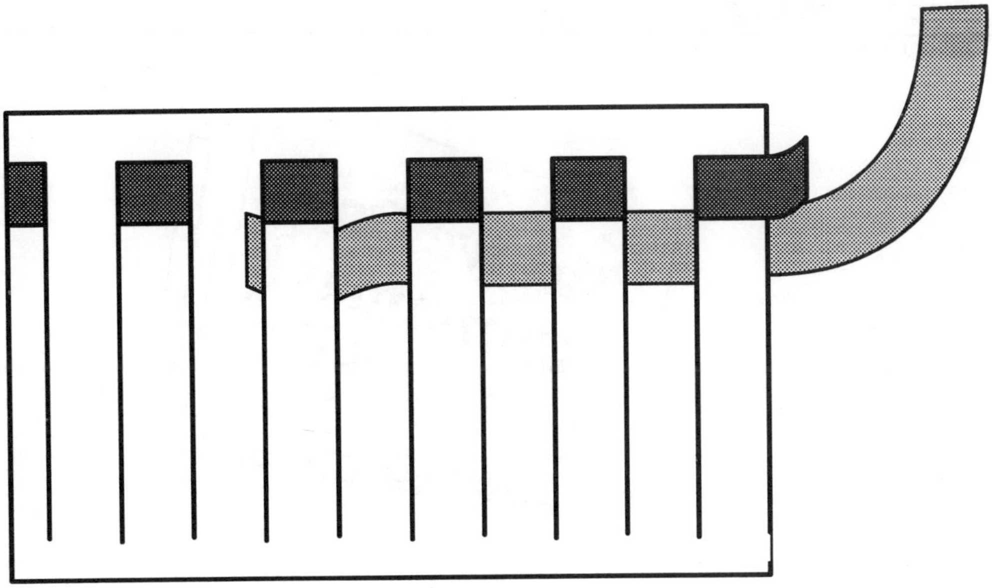

THE TWELVE DAYS OF CHRISTMAS

Description of Booklet (Group Booklet)

Two or more group booklets may be made. The cover consists of two pieces of 12" x 18" oak tag. Inner pages are 12" x 18" sheets of pastel colored construction paper.

Learning Skills Involved

Creativity and eye-hand coordination.

Method of Preparation

Children count off from 1 to 12, then begin again until each one has a number. If there are more than 12 students to a booklet, some may illustrate the cover or two students may illustrate the same page.

The "Twelve Days of Christmas" is printed on the chalkboard. Students illustrate their match number. For example, student # 1 draws a partridge in a pear tree, student # 2 draws two turtle doves, etc.

Each team of 12 tries to create the most beautiful booklet. Paper, pencils, felt-tipped markers, glitter, etc., should be provided on the worktable. Titles should be printed on the bottom of each page.

Finished booklets may be placed on the bookshelf during the holiday season.

Preparation Time

One or two days.

Note: The complete song is given on the following page.

THE TWELVE DAYS OF CHRISTMAS

The first day of Christmas my true love sent to me
A partridge in a pear tree.

The second day of Christmas my true love sent to me
Two turtle doves and a partridge in a pear tree.

The third day of Christmas my true love sent to me
Three French hens, two turtle doves, and a partridge in a pear tree.

The fourth day of Christmas my true love sent to me
Four calling birds, three French hens, two turtle doves, and
A partridge in a pear tree.

The fifth day of Christmas my true love sent to me
Five gold rings, four calling birds, three French hens,
Two turtle doves and a partridge in a pear tree.

The sixth day of Christmas my true love sent to me
Six geese a-laying, five gold rings, four calling birds,
Three French hens, two turtle doves and a partridge in a pear tree.

The seventh day of Christmas my true love sent to me
Seven swans a-swimming, six geese a-laying, five gold rings,
Four calling birds, three French hens, two turtle doves and
A partridge in a pear tree.

The eighth day of Christmas my true love sent to me
Eight maids a-milking, seven swans a-swimming, six geese a-laying,
Five gold rings, four calling birds, three French hens,
Two turtle doves and a partridge in a pear tree.

The ninth day of Christmas my true love sent to me
Nine drummers drumming, eight maids a-milking,
Seven swans a-swimming, six geese a-laying, five golden rings,
Four calling birds, three French hens, two turtle doves and
A partridge in a pear tree.

The tenth day of Christmas my true love sent to me
Ten pipers piping, nine drummers drumming, eight maids a-milking,
Seven swans a-swimming, six geese a-laying, five gold rings,
Four calling birds, three French hens, two turtle dives and
A partridge in a pear tree.

The eleventh day of Christmas my true love sent to me
Eleven ladies dancing, ten pipers piping, nine drummers drumming,
Eight maids a-milking, seven swans a-swimming, six geese a-laying,
Five gold rings, four calling birds, three French hens, two turtle doves
And a partridge in a pear tree.

The twelfth day of Christmas my true love sent to me
Twelve lords a-leaping, eleven ladies dancing, ten pipers piping,
Nine drummers drumming, eight maids a-milking, seven swans a-swimming,
Six geese a-laying, five gold rings, four calling birds, three French hens,
Two turtle doves and a partridge in a pear tree.

NEW YEAR'S RESOLUTIONS

Description of Booklet (Group Booklet)

This booklet consists of two 12" x 18" cover pages of oak tag or posterboard. Inner pages of manila paper are stapled inside the cover when they have been completed by students.

Learning Skills Involved

Eye-hand coordination and imagination.

Method of Preparation

A class discussion of resolutions and promises should precede work.

Each student will write and illustrate a New Year's resolution. Paper, pencils, crayons and/or felt-tipped markers are provided.

Preparation time

One day.

THE EASTER BOOK

Description of Booklet (Group Booklet)

This is an oval-shaped novelty booklet cut from two pieces of 12" x 18" oak tag or colored posterboard. Inner pages consist of manila paper cut to fit the oval shape. The booklet is stapled together after all children have completed their pages.

Learning Skills Involved

Eye-hand coordination and imagination.

Method of Preparation

Each child is given a manila oval and asked to draw an Easter picture. Pencils, crayons and/or felt-tipped markers are placed on the work table. Examples of possible illustrations include:

1. me in my Easter outfit
2. an Easter egg hunt
3. the Easter bunny
4. a basket of Easter eggs or Easter flowers

Preparation Time

One day.

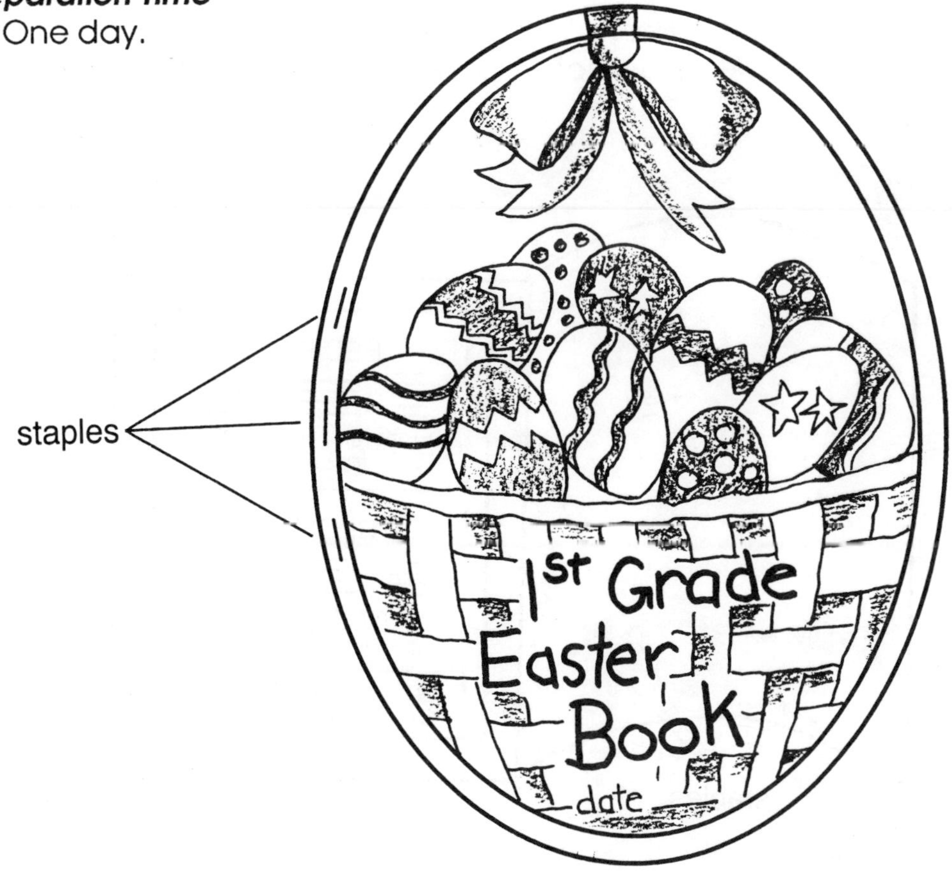

MOTHER'S DAY COOKBOOK

Description of Booklet

This booklet consists of 8 1/2" x 11" pages of duplicated papers. When complete, it is assembled by the teacher who makes a copy for each child in the classroom. Covers may be colored construction paper decorated by students.

Learning Skills Involved

Listening and following directions.

Method of Preparation

Notes are sent home with children asking a parent (or parents) for a favorite recipe to be included in the cookbook.
An alternative method is to record each recipe that has been cooked in class by students during the year.

The teacher types two or three recipes on each page, duplicates them, then staples them together. Pages may contain simple drawings that may be colored.

Preparation Time

Two or three weeks in order to collect all recipes. Each student should be represented. If parents do not respond, the teacher can supply the needed recipe.

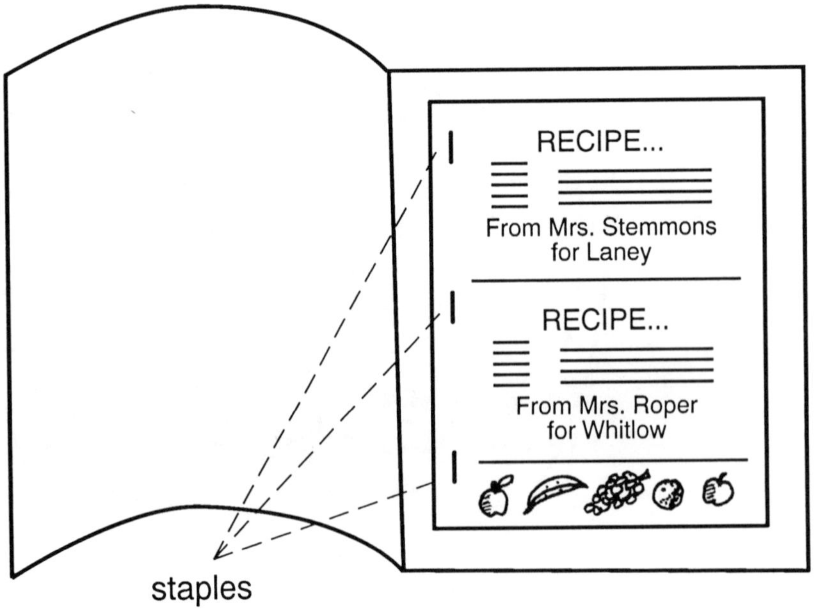

RELIGIONS OF THE WORLD

Description of Booklet

This booklet consists of a 12" x 18" cover of colored construction paper that is folded in half. Inner pages of typing paper are stapled inside the cover.

Learning Skills Involved

Eye-hand coordination, observation, and cognitive discrimination.

Method of Preparation

A class discussion of various religions is held including their main holidays and beliefs.

Stencils of various religious symbols may be offered. Reference books may be placed on the work table. Students are given pencils, felt-tipped markers and crayons. Each page in the booklet should be devoted to one religion. Four or five religions may be selected by students to study.

Preparation Time

Four or more days.

Some facts about five of the world's great religions are given on the following page.

Some Basic Facts about Five Great Religions

Christianity

Christianity is based on the teachings of Jesus Christ as stated in the New Testament.

Important Christian holidays include: Christmas, St. Valentine's Day and Easter.

Christians call their houses of worship churches or cathedrals.

Judaism

Judaism is based on the teachings in the Old Testament.

Important Jewish holidays include: Rosh Hashana (Jewish New Year's), Yom Kippur, Chanukah, and Passover.

Jews call their houses of worship synagogues or temples.

Moslem

Moslems believe in the god Allah and his prophet Mohammed, as stated in the Koran.

Important Moslem holidays include: Milad-un-Nabi (Mohammed's birthday) and Ramadan, the holy month of fasting.

Moslems call their houses of worship mosques.

Hinduism

Hindus worship many gods. They believe that all life is sacred as stated in their three holy books.

Important Hindu festivals include the Hindu New Year, Holi, and the birthday of Krishna and Duvali (the Feast of Lights).

The Hindus call their places of worship shrines or temples.

Buddhism

Buddhists follow the way of one man: Buddha. They believe people should lead a simple way of life.

Important Buddhist festivals include: Wesak (the 3 main events in the life of Buddha) and All Soul's Day (when spirits return to their homes).

Buddhist houses of worship are called shrines, temples, or pagodas.

CREATION

(or In the Beginning)

Description of Booklet

This booklet consists of seven 8" round pages of manila paper. Each page will represent one day. Upon completion, pages are stapled together.

Learning Skills Involved

Eye-hand coordination, listening and imagination.

Method of Preparation

Each day, the teacher and class discuss one aspect of creation. Using pencils, felt-tipped markers and/or crayons, students illustrate the story on their 8" round papers.

According to the story in the Old Testament, the following days are regarded:

Day 1	Night and Day
Day 2	Heavens and Earth
Day 3	Plants and Trees
Day 4	Sun, Moon and Stars
Day 5	Fish and Birds
Day 6	Man and other Animals
Day 7	God rested

Preparation Time

Seven days.

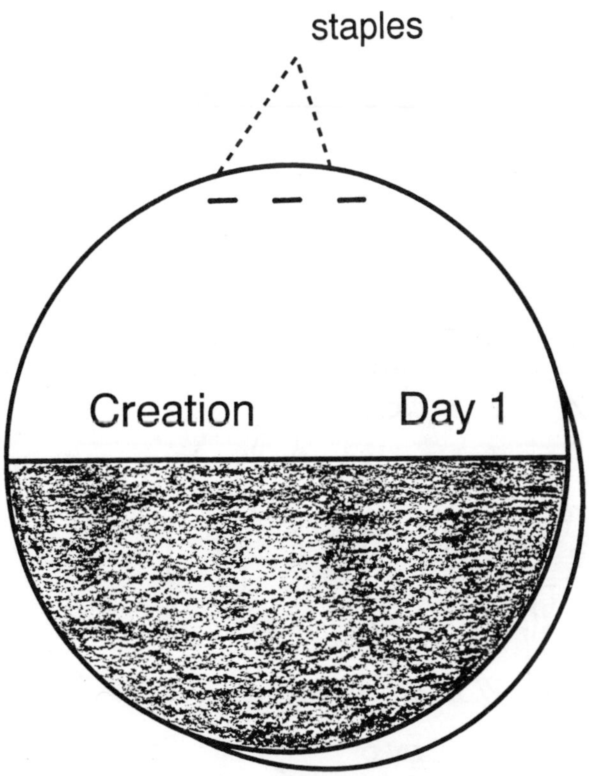

AND GOD CREATED . . .

NOAH'S ARK

Description of Booklet (Group Booklet)

This is a large ark-shaped booklet or a 12" x 18" cover of colored construction paper with an ark glued on the front cover. Inner pages are ark-shaped or 12" x 18" manila paper.

Learning Skills Involved

Eye-hand coordination, listening, following directions, and creativity.

Method of Preparation

The story of Noah is discussed by the class. Students are then given ark-shaped paper or plain manila paper, pencils and crayons. They may draw any part of the story they desire. Suggestions include:

1. Noah – and perhaps his wife, three sons and their wives.
2. pairs of animals
3. the rain
4. the dove with the olive leaf

Teachers or students will title their pages. The teacher will arrange them in chronological order and staple the booklet together. The booklet may be placed on the library shelf for the children to "read."

Preparation Time

One day.

ADDITIONAL SUGGESTIONS FOR BOOKLETS

Vikings
Pirates
Conquerors
Eskimos
Cowboys
Astronauts
Heroes
Famous ball players (or other sports)
Famous dogs (or horses)
Great men in history
Great women in history
Favorite Sesame Street characters (or other TV shows)
Rain forests of the world
Great rivers of the world
Great deserts of the world
National parks
Great mountains or mountain ranges
The circus
Flags of North America (or other continents)
Famous monuments of the world:
    Stonehenge, the pyramids of Egypt and Mexico, the Great Wall of China, the Easter Island Statues, the cave drawings in Spain, etc.
Famous bridges
Famous cathedrals
Book of first aide
Musical Instruments:
    Brass – tuba, trumpet, French horn, trombone, bugle
    Percussion – triangle, drums, cymbals, chimes, xylophone
    Keyboard – piano, harpsichord, organ, accordion
    String – violin, harp, banjo, cello, guitar
    Woodwind – bassoon, clarinet, saxophone, flutes, oboe
Famous persons in your state